Y0-DCU-873

PLEASE STAMP DATE DUE, BOTH BELOW AND ON CARD

DATE DUE	DATE DUE	DATE DUE	DATE DUE

PLEASE DO NOT BORROW
UNTIL ____ DEC 29 '03

GL-15

Millikan
QC462.H1 P74 2003
Precision physics of simple
atomic systems

Lecture Notes in Physics

Editorial Board

R. Beig, Wien, Austria
B.-G. Englert, Ismaning, Germany
U. Frisch, Nice, France
P. Hänggi, Augsburg, Germany
K. Hepp, Zürich, Switzerland
W. Hillebrandt, Garching, Germany
D. Imboden, Zürich, Switzerland
R. L. Jaffe, Cambridge, MA, USA
R. Lipowsky, Golm, Germany
H. v. Löhneysen, Karlsruhe, Germany
I. Ojima, Kyoto, Japan
D. Sornette, Nice, France, and Los Angeles, CA, USA
S. Theisen, Golm, Germany
W. Weise, Trento, Italy, and Garching, Germany
J. Wess, München, Germany
J. Zittartz, Köln, Germany

Springer
Berlin
Heidelberg
New York
Hong Kong
London
Milan
Paris
Tokyo

Physics and Astronomy

ONLINE LIBRARY

http://www.springer.de/phys/

The Editorial Policy for Edited Volumes

The series *Lecture Notes in Physics* (LNP), founded in 1969, reports new developments in physics research and teaching - quickly, informally but with a high degree of quality. Manuscripts to be considered for publication are topical volumes consisting of a limited number of contributions, carefully edited and closely related to each other. Each contribution should contain at least partly original and previously unpublished material, be written in a clear, pedagogical style and aimed at a broader readership, especially graduate students and nonspecialist researchers wishing to familiarize themselves with the topic concerned. For this reason, traditional proceedings cannot be considered for this series though volumes to appear in this series are often based on material presented at conferences, workshops and schools.

Acceptance

A project can only be accepted tentatively for publication, by both the editorial board and the publisher, following thorough examination of the material submitted. The book proposal sent to the publisher should consist at least of a preliminary table of contents outlining the structure of the book together with abstracts of all contributions to be included. Final acceptance is issued by the series editor in charge, in consultation with the publisher, only after receiving the complete manuscript. Final acceptance, possibly requiring minor corrections, usually follows the tentative acceptance unless the final manuscript differs significantly from expectations (project outline). In particular, the series editors are entitled to reject individual contributions if they do not meet the high quality standards of this series. The final manuscript must be ready to print, and should include both an informative introduction and a sufficiently detailed subject index.

Contractual Aspects

Publication in LNP is free of charge. There is no formal contract, no royalties are paid, and no bulk orders are required, although special discounts are offered in this case. The volume editors receive jointly 30 free copies for their personal use and are entitled, as are the contributing authors, to purchase Springer books at a reduced rate. The publisher secures the copyright for each volume. As a rule, no reprints of individual contributions can be supplied.

Manuscript Submission

The manuscript in its final and approved version must be submitted in ready to print form. The corresponding electronic source files are also required for the production process, in particular the online version. Technical assistance in compiling the final manuscript can be provided by the publisher's production editor(s), especially with regard to the publisher's own LaTeX macro package which has been specially designed for this series.

LNP Homepage (http://www.springerlink.com/series/lnp/)

On the LNP homepage you will find:
−The LNP online archive. It contains the full texts (PDF) of all volumes published since 2000. Abstracts, table of contents and prefaces are accessible free of charge to everyone. Information about the availability of printed volumes can be obtained.
−The subscription information. The online archive is free of charge to all subscribers of the printed volumes.
−The editorial contacts, with respect to both scientific and technical matters.
−The author's / editor's instructions.

S.G. Karshenboim V.B. Smirnov (Eds.)

Precision Physics of Simple Atomic Systems

Springer

Editors

Savely G. Karshenboim
D.I. Mendeleev Institute for Metrology
198005 St. Petersburg
Russia
and
Max Planck Institute for Quantum Optics
85748 Garching
Germany

Valery B. Smirnov
Russian Center for Laser Physics
St. Petersburg State University
198504 St. Petersburg
Russia

Cataloging-in-Publication Data applied for

A catalog record for this book is available from the Library of Congress.

Bibliographic information published by Die Deutsche Bibliothek
Die Deutsche Bibliothek lists this publication in the Deutsche Nationalbibliografie;
detailed bibliographic data is available in the Internet at http://dnb.ddb.de

ISSN 0075-8450
ISBN 3-540-40489-9 Springer-Verlag Berlin Heidelberg New York

This work is subject to copyright. All rights are reserved, whether the whole or part of the
material is concerned, specifically the rights of translation, reprinting, reuse of illustra-
tions, recitation, broadcasting, reproduction on microfilm or in any other way, and
storage in data banks. Duplication of this publication or parts thereof is permitted only
under the provisions of the German Copyright Law of September 9, 1965, in its current
version, and permission for use must always be obtained from Springer-Verlag. Violations
are liable for prosecution under the German Copyright Law.

Springer-Verlag Berlin Heidelberg New York
a member of BertelsmannSpringer Science+Business Media GmbH

http://www.springer.de

© Springer-Verlag Berlin Heidelberg 2003
Printed in Germany

The use of general descriptive names, registered names, trademarks, etc. in this publication
does not imply, even in the absence of a specific statement, that such names are exempt
from the relevant protective laws and regulations and therefore free for general use.

Typesetting: Camera-ready by the authors/editor
Cover design: *design & production*, Heidelberg

Printed on acid-free paper
54/3141/du - 5 4 3 2 1 0

Foreword

Fashions come and go in physics as in any other human endeavor. However, the simple hydrogen atom has captured the attention of physicists for more than a century, and this book gives testimony of the continuing vigor, promise, and fascination held by the physics of simple atomic systems.

During the first three decades of the 20th century, the hydrogen atom has provided the Rosetta stone for deciphering the mysteries of quantum physics. Spectroscopy of the regular hydrogen Balmer spectrum has inspired numerous pathbreaking discoveries, from the Bohr model and the old quantum physics of Sommerfeld to the wave mechanics of de Broglie and Schrödinger and on to the relativistic quantum theory of Dirac. The line profile of the red Balmer line also gave the first hints for a level shift due to vacuum fluctuations. Willis Lamb has been awarded the Nobel Prize for confirming this shift after the second world war, laying one of the main experimental foundations of modern quantum electrodynamics.

The interest in optical spectroscopy of hydrogen revived in the 1970s, after highly monochromatic tunable lasers became available which could eliminate the Doppler broadening of spectral lines by nonlinear laser spectroscopy. For the first time, single fine structure components could be resolved in the hydrogen Balmer lines. Even the first laser measurement of the wavelength of the red Balmer line by saturation spectroscopy in my former group at Stanford yielded a tenfold improvement in the accuracy of the Rydberg constant, one of the cornerstones in the system of fundamental constants. Since then several experimental groups have pursued the tantalizing goal to test fundamental theory and to measure important constants by ever more precise laser spectroscopy of atomic hydrogen. This quest has inspired numerous advances in spectroscopic techniques and instrumentation. Methods such as polarization spectroscopy, Doppler-free two-photon spectroscopy, and even laser cooling have been invented and refined during th is pursuit since the 1970s.

In the 1990, the achieved spectral resolution began to challenge the accuracy limits of optical wavelength metrology and motivated an intense search for better techniques to measure the frequency of light. In the past few years, this quest has culminated in the invention of the optical frequency comb based on femtosecond laser, which provides a direct link between optical and microwave frequencies and which permits the comparison of different frequencies with unprecedented levels of precision. In the first measurement of an absolute optical frequency with the help of such a comb synthesizer at the Max-Planck-Institut für Quantenoptik in Garching during the summer of 1999, the frequency of the sharp hydrogen 1S-2S two-photon transition has been compared to the microwave frequency of a transportable cesium fountain clock to within 1.9 parts in 10^{14}. Together with a new such measurement in February of 2003, this experiments is setting new limits on a possible slow variation of the fine structure constant with the evolution of the universe. The accuracy of such spectroscopic measurements is now limited by the microwave atomic clocks. However, the same spectroscopic methods are enabling the development of novel optical atomic clocks which are expected to push the limits of spectroscopic precision to parts in 10^{18} within the next decade.

Such prospects have inspired renewed interest in the quantum electrodynamic theory of atomic hydrogen levels. Theorists in atomic physics and in particle physics are discovering problems of common interest and are beginning to share their insights, tools, and methods. New approaches, new algorithms, and new computer power are being harnessed to conquer previously elusive higher order corrections.

In the past, comparisons of theory and experiment have been hampered by our poor knowledge of the quadratic charge radius of the proton. However, a laser measurement of the Lamb shift of muonic hydrogen, now well underway at the Paul Scherrer Institute, should finally yield a precise measurement of the proton size. This study of the muonic atom is an example for the growing list of simple atoms under study. The experimental frontier continues to expand in the direction of hydrogen-like short-lived exotic atoms and heavier ions. Another challenging experiment will be precision laser spectroscopy of antihydrogen. The first slow antihydrogen atoms have now been produced by the ATHENA and ATRAP teams at CERN, and future experiments may well reveal conceivable differences between matter and antimatter.

Considering such prospects it seems quite likely that atomic hydrogen will again become a Rosetta stone for deciphering the secrets of nature. Perhaps the biggest surprise in this continuing endeavor would be if we found no surprises.

The 2002 conference on the physics of simple atoms brought together many scientists working at this exciting frontier. Fortunately the most important contributions to the meeting are published in this book and are thus now available to a wider audience.

Munich,   *Theodor W. Hänsch*
April, 2003

Preface

Because of their apparent "simplicity" simple atoms present a great challenge and temptation to experts in various branches of physics from fundamental problems of particle physics to astrophysics, applied physics and metrology. This book is based on the presentations at the International Conference on *Precision Physics of Simple Atomic Systems* (PSAS 2002) whose primary target was to provide an effective exchange between physicists from different fields.

From the early days of modern physics, studies of simple atoms involved basic ideas which have essentially contributed to the creation of the present day physical picture based on the Schrödinger and Dirac theories, quantum electrodynamics (QED) with the renormalization approach etc. Today, the precision physics of simple atoms involves the most sophisticated experimental and theoretical methods and plays an important role in the progress of laser, atomic, nuclear and particle physics and metrology, delivering the most accurate available data.

This book covers a broad range of problems related to simple atoms and precision measurements: spectroscopy of hydrogen, helium, muonic and exotic atoms, highly charged ions, nuclear structure and its effects on atomic energy levels, highly accurate determination of values of fundamental physical constants and the search for their possible variation with time, QED tests and precision mass spectroscopy.

This is the second book on the subject within a relatively short time, following *Hydrogen Atom: Precision Physics of Simple Atomic Systems* published by Springer in 2001 (LNP Vol. 570). However, the collection of reviews presented here has no essential overlap with the previous volume.

Simple atoms play an important role in science teaching, offering a good demonstration system to apply quantum mechanics. We include in the book two lectures on the theory of Coulomb atomic systems. These two tutorial papers on hydrogen-like atoms present a collection of applications to actual problems of advanced quantum theory of the hydrogen atom and other hydrogen-like systems.

This book presents the state-of-the-art and recent progress in studies of simple atoms and related questions. We also tried to cover topics missing in the former book and we hope we succeeded in that. The present book is prepared following the review presentations at *PSAS 2002* which took place in St. Petersburg on June, 30–July, 4, 2002. More detail about the meeting can be found at http://psas2002.vniim.ru. Several selected progress reports are also included.

Support from the Russian Foundation for Basic Research, Russian Center of Laser Physics, D.I. Mendeleev Institute for Metrology (VNIIM), Max-Planck Institut für Quantenoptik is gratefully acknowledged by the Organizing Committee.

We would like specially to express our gratitude to Gordon Drake and the Canadian Journal of Physics, published by the NRC Research Press, and to Springer-Verlag for the agreement to publish in this book an enlarged version of four papers presented in the conference issue of the Canadian Journal of Physics. In particular we gratefully acknowledge that several of the pictures used by our contributors have been already published by them in the Canadian Journal of Physics.

To my great sorrow, Prof Valery Smirnov, the head of the Russian Center of Laser Physics and a co-chairman of the Conference died recently and will not see our book published. To him we also owe our gratitude.

St.Petersburg and Garching   *Savely Karshenboim*
April, 2003

Contents

Recent Progress with Precision Physics of Simple Atoms
S.G. Karshenboim, V.B. Smirnov 1
1 Introduction 1
2 Recent Progress in the Study of Hydrogen and Helium 2
3 Progress in the Study of Muonium and Positronium 3
4 Progress in the Precision Study of Highly Charged Ions 3
5 Advances in Determination of Fundamental Constants 5
6 New Results on Precision Tests of Quantum Electrodynamics 6
7 Search for Variations of the Fundamental Constants 6
8 Study of Muonic and Exotic Atoms 7
9 Quantum Mechanics of Hydrogen-Like Atoms: Tutorial 8
10 About PSAS 2002 Conference 9
References 10

Part I   The Hydrogen Atom

Coulomb Green Function and Its Applications in Atomic Theory
L.N. Labzowsky, D.A. Solovyev 15
1 Coulomb Green Function for the Schrödinger Equation 15
2 Sturmian Expansion 18
3 Two-Photon Decay of Atomic Levels 20
4 Lamb Shift in the Hydrogen Atom 23
5 Nonresonant Corrections in Atomic Hydrogen 25
6 Relativistic Coulomb Green Function 28
7 Relativistic Polarizability of the H-Like Ions 32
References 34

Part II   Muonic and Exotic Atoms and Nuclear Effects

**Atomic Cascade and Precision Physics with Light Muonic
and Hadronic Atoms**
T.S. Jensen, V.E. Markushin 37
1 Introduction 37
2 Atomic Cascade 38

3   Muonic Hydrogen 42
4   Pionic Hydrogen 46
5   Kaonic Hydrogen 49
6   Antiprotonic Hydrogen 52
7   Conclusion 54
References 56

The Structure of Light Nuclei
and Its Effect on Precise Atomic Measurements
J.L. Friar 59
1   Introduction 59
2   Myths of Nuclear Physics 60
3   The Nuclear Force 60
4   Calculations of Light Nuclei 65
5   What Nuclear Physics Can Do for Atomic Physics 67
6   The Proton Size 73
7   What Atomic Physics Can Do for Nuclear Physics 74
8   Summary and Conclusions 76
References 76

Deeply Bound Pionic States as an Indicator of Chiral
Symmetry Restoration
T. Yamazaki 81
1   Pionic Atoms – Old History and New Frontier 81
2   Prediction for Quasi-Stable Pionic Nuclei 82
3   Observation of Deeply Bound Pionic States 82
4   Pion-Nucleus Interaction 84
5   Evidence for In-Medium Restoration of Chiral Symmetry 89
References 91

Part III   Hydrogen-Like Ions

Virial Relations for the Dirac Equation and Their Applications
to Calculations of Hydrogen-Like Atoms
V.M. Shabaev 97
1   Introduction 97
2   Derivation of the Virial Relations for the Dirac Equation 98
3   Application of the Virial Relations for Evaluation
     of the Average Values 99
4   Application of the Virial Relations for Calculations
     of Higher-Order Corrections 101
5   Calculations of the Bound-Electron g Factor and the Hyperfine
     Splitting in H-Like Atoms 105
6   Other Applications of the Virial Relations 111
7   Conclusion 112
References 113

Lamb Shift Experiments on High-Z One-Electron Systems
T. Stöhlker, D. Banaś, H. Beyer, A. Gumberidze 115
1   Introduction 115
2   The Storage Ring ESR 118
3   X-Ray Spectroscopy at the ESR 120
4   Summary and Outlook 135
References 135

Part IV   Testing Quantum Electrodynamics

**Simple Atoms, Quantum Electrodynamics,
and Fundamental Constants**
S.G. Karshenboim 141
1   Introduction 141
2   Rydberg Constant and Lamb Shift in Hydrogen 142
3   Hyperfine Structure and Nuclear Effects 145
4   Hyperfine Structure of the $2s$ State in Hydrogen, Deuterium, and
    Helium-3 Ion 147
5   Hyperfine Structure in Muonium and Positronium 148
6   g Factor of Bound Electron and Muon in Muonium 150
7   g Factor of a Bound Electron in a Hydrogen-Like Ion with Spinless
    Nucleus 152
8   The Fine Structure Constant 155
9   Summary 157
References 158

**Resent Results and Current Status of the Muon (g–2)
Experiment at BNL**
*S.I. Redin, G.W. Bennett, B. Bousquet, H.N. Brown, G. Bunce,
R.M. Carey, P. Cushman, G.T. Danby, P.T. Debevec, M. Deile, H. Deng,
W. Deninger, S.K. Dhawan, V.P. Druzhinin, L. Duong, E. Efstathiadis,
F.J.M. Farley, G.V. Fedotovich, S. Giron, F. Gray, D. Grigoriev,
M. Grosse-Perdekamp, A. Grossmann, M.F. Hare, D.W. Hertzog, X. Huang,
V.W. Hughes, M. Iwasaki, K. Jungmann, D. Kawall, M. Kawamura,
B.I. Khazin, J. Kindem, F. Krinen, I. Kronkvist, A. Lam, R. Larsen,
Y.Y. Lee, I.B. Logashenko, R. McNabb, W. Meng, J. Mi, J.P. Miller,
W.M. Morse, D. Nikas, C.J.G. Onderwater, Yu.F. Orlov, C. Ozben,
J. Paley, Q. Peng, J. Pretz, R. Prigl, G. zu Putlitz, T. Qian, O. Rind,
B.L. Roberts, N.M. Ryskulov, P. Shagin, S. Sedykh, Y.K. Semertzidis,
Yu.M. Shatunov, E.P. Solodov, E.P. Sichtermann, M. Sossong,
A. Steinmetz, L.R. Sulak, C. Timmermans, A. Trofimov, D. Urner,
P. von Walter, D. Warburton, D. Winn, A. Yamamoto, D. Zimmerman* . . 163
1   Introduction 164
2   Muon (g–2) Experiment E821 at BNL 165
3   Magnetic Field Measurement and Control 167

4 Data Analysis and Results 169
5 Standard Model Prediction for a_μ 173
6 Outlook 174
References 174

Part V   Precision Measurements and Fundamental Constants

Single Ion Mass Spectrometry at 100 ppt and Beyond
S. Rainville, J.K. Thompson, D.E. Pritchard 177

1 Overview 177
2 Scientific Applications 178
3 Experimental Techniques 180
4 Simultaneous Measurements 187
5 Subthermal Detection 193
6 Conclusion 196
References 196

Current Status of the Problem of Cosmological Variability of Fundamental Physical Constants
D.A. Varshalovich, A.V. Ivanchik, A.V. Orlov, A.Y. Potekhin,
P. Petitjean 199

1 Introduction... 199
2 Tests for Possible Variations of Fundamental Constants 200
3 Conclusions 207
References 208

Appendix:
Proceedings of International Conference on Precision Physics of Simple Atomic Systems (St. Petersburg, 2002) – Table of Contents
Canadian Journal of Physics **80**(11) (2002) 211

Index 215

List of Contributing Participants

James L. Friar, Los Alamos National Laboratory, Los Alamos, NM, USA
friar@lanl.gov

Savely G. Karshenboim, D.I.Mendeleev Institute for Metrology, 198005
St. Petersburg, Russia, and Max-Planck-Institut für Quantenoptik, D-85748
Garching, Germany, sek@mpq.mpg.de

Leonti Labzowsky, St. Petersburg State University, 198904 St. Petersburg,
Russia, leonti@landau.phys.spbu.ru

Valeri E. Markushin, Paul-Scherrer-Institut, CH-5232 Villigen, Switzerland,
valeri.markushin@psi.ch

Simon Rainville, Massachusetts Institute of Technology, Cambridge, MA
02139, USA, simonr@mit.edu

Sergei I. Redin, Budker Institute for Nuclear Physics, Novosibirsk, Russia
630090, redin@inp.nsk.su

Vladimir M. Shabaev, St. Petersburg State University, 198904 St. Petersburg,
Russia, shabaev@pobox.spbu.ru

Valery B. Smirnov, Russian Center of Laser Physics St. Petersburg State
University 198504 St. Petersburg, Russia, vbs@home.rclph.spbu.ru

Thomas Stöhlker, Gesellschaft für Schwerionenforschung, 64291 Darmstadt,
Germany, T.Stoehlker@gsi.de

Dmitri A. Varshalovich, Ioffe Physical-Technical Institute, St.-Petersburg,
194021, Russia, varsh@astro.ioffe.rssi.ru

Toshimitsu Yamazaki, RIKEN, Wako-shi, Saitama-ken, 351-0198 Japan,
yamazaki@nucl.phys.s.u-tokyo.ac.jp

Recent Progress in the Precision Physics of Simple Atoms

Savely G. Karshenboim[1,2] and Valery B. Smirnov[3]

[1] D.I. Mendeleev Institute for Metrology (VNIIM), 198005 St. Petersburg, Russia
[2] Max-Planck-Institut für Quantenoptik, 85748 Garching, Germany
[3] Russian Center of Laser Physics, St. Petersburg State University, 198504
   St. Petersburg, Russia

Abstract. An introduction into the recent progress in precision physics of simple atoms is presented. Special attention is paid to the review contributions presented in this book.

1 Introduction

Precision physics of simple atom is a field which involves scientists from different part of physics. Only a short list of contributions from those parts includes:

- *Laser physics* offers high-resolution spectroscopic methods in the optical domain which can now lead to accurate results even for traditionally microwave effects.
- *Microwave spectroscopy* still delivers new accurate data and especially for properties of particles and ions at a homogenous magnetic field.
- *Atomic physics* with *trapping* of simple atoms and particles supplies us with a powerful tool for precision measurements.
- Progress in the *frequency metrology* offers a good chance to look for a possible *time variation* of values of the fundamental constants which can give a limitation on unification theories and is helpful to realize a new generation of the frequency standards.
- Receiving high precision experimental data, *quantum electrodynamics* (QED) successfully produces theoretical predictions and develops efficient methods for high-order perturbative corrections and so-called "exact" calculations.
- Those "exact" calculations (without any expansion over the Coulomb strength $Z\alpha$) are important even for $Z = 1$ (hydrogen and muonium). Applying accelerators, the data on QED effects (such as the Lamb shift and corrections to the hyperfine interval and electron magnetic moment) are now available for a broad range of *highly charged ion* up to hydrogen-like uranium.
- Due to extremely high accuracy of theory and experiments, their comparison delivers us the crucial data on proton structure and *structure* of light *nucleus*.
- *Accelerator physics* allows to form *muonic* and *exotic* atoms which offer us even the most impressive access to particle properties.
- Highly accurate determination of values of the *fundamental constants* provides us with a promising way on reproduction of *natural units* and is helpful presently for reproduction of ohm and volt, based on macroscopic quantum effects.

All these mentioned items and a number of not mentioned ones are related to a broad range of the problems which are simoltanously separated and correlated in a sense. They are essentially separated because, due to a high accuracy of theory and experiment and deeply advanced development of their methods, most scientists used to be greatly concentrated on their particular problems lying in significantly different branches of physics. However, their work would have no sense if their results could not find applications beyond their subfield. Such a pattern involving the researches from the subfields, which looks almost completely independent, but with a crucial overlap of their results and their applications, is the very soul of precision physics of simple atoms.

Study of simple atoms is also of a methodological interest since theory of hydrogen and hydrogen-like atoms presents the most accurate and most advanced theory of a particular quantum object. Basic text books on quantum mechanics used to have a chapter or so on hydrogen and other simple atoms. However, there are only a few books completely devoted to physics of simple atoms [1–4] and our book is to be an addition to that collection. The present book is based on presentations at the International conference on *Precision Physics of Simple Atomic Systems* (PSAS 2002) and it follows the book [4] which presented the reviews and contributed papers of the International conference *Hydrogen atom: Precision Physics of Simple Atomic Systems* (2000). The book is intended to summarize the state of the art in the field paying special attention to recent progress since the former publication [4] (see also [5]) and to a few topics missed there.

This time the reviews and the contributed papers are published separately. The latter are collected in the conference issue of the Canadian Journal of Physics **80**(11) (2002) and its table of contents can be found in the Appendix to this book [6].

In this introductory paper we briefly summarize recent progress in the field. We overview briefly controbutions to this book and to the conference issue and also consider a continuation of the study presented in the previous book.

2   Recent Progress in the Study of Hydrogen and Helium

Precision spectroscopy was reviewed in book [4] for the hydrogen atom in [7] and for helium in [8] and despite dramatic advances by time of the former conference in 2000, exciting progress in the field is still possible. Hydrogen optical spectroscopy had achieved great results in recent times and a demonstration of its power was a successful measurement of the hydrogen Lamb shift, a value traditionally measured by microwave means. Another invasion into the microwave domain was presented at PSAS 2002. It is related to the hyperfine splitting of the metastable 2s state. A theory of 2s hyperfine structure was considered in [9] and it was shown there that there were still some problems to be solved. The theory was improved later [10,11]. An attractive aspect of 2s hyperfine structure is a significant cancellation of the nuclear effects that occurs when combined with the 1s hyperfine interval

$$D_{21} = 8 \cdot E_{\text{HFS}}(2s) - E_{\text{HFS}}(1s) \,. \qquad (1)$$

This essential cancellation opens the possibility for a precision test of quantum electrodynamics applied to bound state problems (so-called *bound state QED*). The theoretical study [9,11] motivated an experiment [12,13] which reached an accuracy superseding the microwave methods [14]. The hyperfine splitting of the 2s state was measured as a difference of two optical $1s - 2s$ transitions (two-photon Doppler-free excitation) for states with the atomic angular momentum $F = 0$ and $F = 1$. The uncertainty of the hyperfine splitting was found to be a few parts in 10^{15} of the "big" $1s - 2s$ interval. A consideration of the difference D_{21} can be found in this book [15].

   In the case of the spectrum of the helium atom, we note a recent breakthrough in theory [16,17] of the fine structure of helium has yielded a new value of the fine structure constant

$$\alpha^{-1} = 137.035\,989\,3(23) \,. \qquad (2)$$

Despite a significant improvement of theory, two major issues remains unsolved: a scattering of experimental data [18–22] and a rather bad agreement between theory and experiment for the "small" splitting between $1s2p\,^3P_2$ and $1s2p\,^3P_1$ (so-called ν_{21}; see Fig. 1 for detail). Both theory and experiment need better understanding of their uncertainties but obviously the situation looks much more promising than a few years ago.

3   Progress in the Study of Muonium and Positronium

Other effective tests of bound state QED can be realized with pure leptonic atoms (muonium and positronium). Since publishing of reviews on their study [23] and [24] only minor progress in the field has been seen. A number of experimental studies of exotic decay modes of positronium was presented [25], while in the case of muonium theoretical results on the g factor of a bound electron and muon [26] and on hadronic contributions to the hyperfine interval were reported [27]. All these results are important as a preliminary step for further progress. Study of the exotic modes of orthopositronium decay is helpful to develop new efficient positronium sources, while in the case of muonium the theoretical progress [27,26] is helpful in clarifying limitations on future bound state QED tests.

4   Progress in the Precision Study of Highly Charged Ions

In light atoms, the electron is bound by the nuclear Coulomb field and has potential energy which is of order of $(Z\alpha)^2 \cdot m_e c^2$. Such an energy is in a sense a perturbation since it is a few orders of magnitude below the relativistic rest mass energy $m_e c^2$. The physics of the hydrogen and other light atoms, i.e. of atoms with the "weakly" bound electrons $((Z\alpha) \ll 1)$, stands somewhere in between

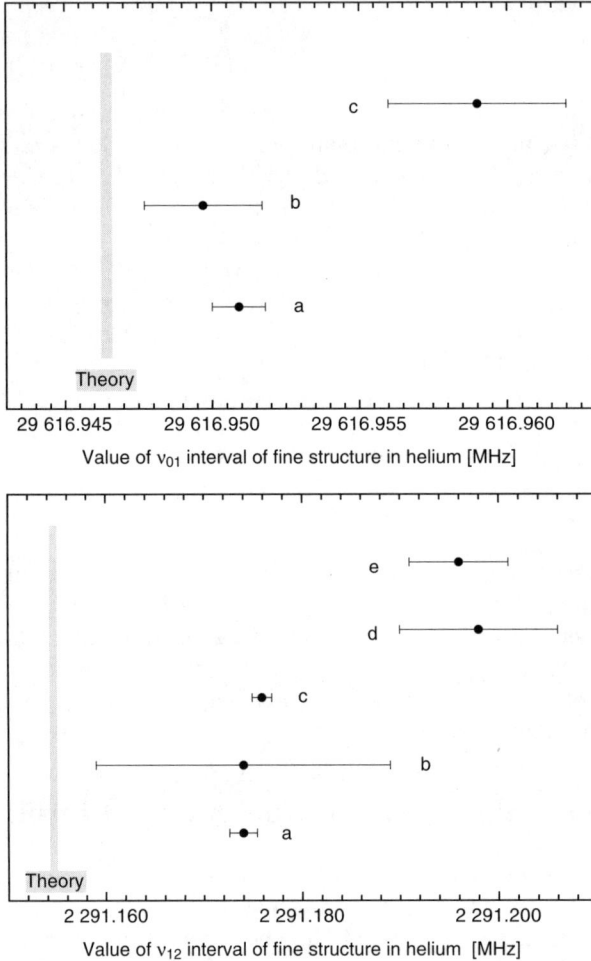

Fig. 1. Fine structure in neutral helium. ν_{01} stands for $1s2p\,^3P_0 - ^3P_1$ and ν_{21} for $1s2p\,^3P_1 - ^3P_2$. The theory is presented accordingly to [17], the experimental data are taken a from [18], b from [19], c from [20], d from [21] and e from [22]. Less accurate data are not included.

two big problems: the QED of free particles and the QED of strongly coupled electrons. The latter can be studied in detail with the help of highly charged ions since the potential energy scales with the value of the nuclear charge Z as Z^2. A specific reason to study highly charged ions, i.e. atoms with most of their electrons stripped, is that in an atomic system with a large nuclear charge Z and only a few electrons, the electron-electron interactions can be treated as a perturbation $(1/Z)$ and thus the ion can be calculated *ab initio*.

The physics of highly charged ions was presented in book [4] in two reviews. One was devoted to spectroscopy of medium-Z ions [28], while the other was related to the g factor of a bound electron in the hydrogen-like carbon [29]. Due to technical reasons the third review on high-Z ions scheduled for [4] was not ready in time and could not contribute to the former book. Now this review is updated for the recent progress [30], presenting in detail recent measurements of the Lamb shift in highly charged hydrogen-like ions. Two former reviews were devoted more to medium Z ions, i.e. ions where $Z\alpha$ is still essentially smaller than unity, while Z is significantly bigger than unity. The latter review is essentially devoted to the hydrogen-like uranium ($Z = 92$), i.e. a high-Z ion.

There has been major progress in both theory and experiment for the g factor of a bound electron. Following the suggestion of [31,5], the electron mass

$$m_e = 0.000\,548\,579\,909\,2(4) \text{ u} \tag{3}$$

and the proton-to-electron mass ratio

$$\frac{m_p}{m_e} = 1836.152\,673\,4(13) \ . \tag{4}$$

were determined [32]. Later the g factor of a bound electron in hydrogen-like oxygen was measured [33] and calculated theoretically to all orders in $Z\alpha$ at the one-loop level [34] (see also [35]) for various ions. Another important achievement of the medium Z physics was an accurate measurement of the intercombination $1s2s\,^1S_0 - 1s2p\,^3P_1$ interval of helium-like silicon [36]. Studies of medium-Z few electron atoms could be helpful in understanding better electron-electron interactions and improving the theory of the helium energy levels (see Sect. 2).

5   Advances in Determination of Fundamental Constants

Totest QED successfully, we have to be aware of accurate values of a few basic fundamental constants which unavoidably enter QED calculations. Part of the recent progress for fundamental constants has been already discussed above. A discussion of the advances in an accurate determination of values of the basic physical constants such as the fine structure constant α is timely because of the next coming adjustment of the fundamental constants, a procedure which is supposed to check the overall consistency of all our data for the fundamental constants and eventually to deliver the most reliable and precise values. The former adjustment performed by CODATA was sketched in [39]. The deadline for the coming one is the end of 2002 and we believe that some of the results considered in the previous book and in this book will contribute to the new recommended values of α, the electron mass, the proton-to-electron mass ratio etc. Applications of QED to determination of the fundamental constants are considered in [15].

Important progress with the fine structure constant was achieved by *Chu and collaborators* who determined the value of h/M_{Cs} by measuring the photon recoil via atomic interferometry [37]. To determine α from h/M_{Cs} one needs to know the cesium mass M_{Cs} and electron mass m_e in atomic mass units (or in the units of the proton mass m_p) and a value of the Rydberg constant Ry:

$$\alpha = \left(\frac{2Ry}{c} \frac{h}{M_{Cs}} \frac{M_{Cs}}{m_p} \frac{m_p}{m_e} \right)^{1/2} . \tag{5}$$

A determination of the Rydberg constant was considered in the former book [7], while measurements of the atomic mass of cesium [38] and of the electron [15] are in part discussed here. A highly accurate measurement of other atomic masses [38] can be used in the future to determine the fine structure constant via the photon recoil and atomic interferometry of other atoms.

6   New Results on Precision Tests of Quantum Electrodynamics

As we mention above, QED calculations can be split into two kinds: quantum electrodynamics of free particles and bound state QED. The bound state calculations reviewed in [15] differ cruicially from the free QED theory, which was presented two years ago in [40] with a theoretical review on calculations of the anomalous magnetic moment of the electron and muon and a new accurate measurement of the anomalous magnetic moment and its consequences is considered here [41]. Measuring the anomalous magnetic moment of the muon with an uncertainty below 1 ppm level opens a new page of the competition between theory and experiment. In contrast to the situation reviewed in 2001, experiment provides now a real test of theory. However, we have to note that although the QED part of theory is known with great accuracy there remain serios limitations due to the uncertainty in the hadronic contribution. Some discrepancy between theory and experiment provoced speculations on possible new physics contributing to the anomalous magnetic moment. A more reasonable target in our mind would be rather a study of the uncertainty of hadronic effects which seems to be not clear enough [42]. Systematic checking of the QED calculations of hundreds of four loop diagrams for the anomalous magnetic moment of the electron and muon is also on the way and some corrections are expected [43].

7   Search for Variations of the Fundamental Constants

For a while high-energy physicists have been looking for different extensions of the Standard Model, for various Grand Unification Theories *etc.* One of the options in looking for consequences for new physics coming from these extension and unification schemes is to search for a possible time (and space) variation of values of the fundamental physical constants. In 2001 the current situation was presented in a progress report by the group of Flambaum and Webb from Sydney [44]. In the present book a detailed review summarizes the state of the art in the field from the point of view of their competitors from the Varshalovich group [45]. An exciting problem arises in the interpretation of astrophysical data, which have shown an inconsistency of available data and their interpretation in terms of relativistic many body atomic theory assuming a "constant" value of the fine structure constant [44,46]. They are related to a possible drift of the fine structure constant as small as $0.6 \cdot 10^{-15}$ yr^{-1} in fractional units.

Recent dramatic progress in frequency spectroscopy reviewed [47] in the former book essentially shifted laboratory activity in the development of new frequency standards towards optical transitions. The most recent results on precision measurements with uncertainties below a part of 10^{14} [48–50,12] allow us to hope that within a few years optical limitations on a possible time variation of the fine structure constant will reach the level of a few parts in 10^{15} per a year. A comparison of two microwave transitions related to the hyperfine splitting in rubidium and cesium performed with ultracold atomic fountains [51] provides us with a promising data on the limitation of a possible time variation of the g factor of the proton [52].

8   Study of Muonic and Exotic Atoms

While studies of the highly charged ions are frequently considered as a way to study QED in strong electric fields (see e.g. [30]), an even stronger field can be reached just by switching from an orbiting electron to a heavier particle, e.g. a muon. The muonic atoms provide a strong field together with a weak potential energy in comparison with $m_\mu c^2$. The "strongness" of the field leads to an enhancement of the production of virtual electron-positron pairs and details of the structure of the spectrum in muonic atoms differ from those for the conventional (electronic) atoms. The review [53] provides an adequate coverage of the subject. We note, that due to reasons of particle physics, few projects on intensive muon sources are under serious consideration [54] and we anticipate a kind of a renaissance of physics of the muonic atoms and the review presented here seems to be timely.

One of reasons to study spectroscopy of muonic atoms is to probe nuclear structure. The muon orbit lies much closer to the nucleus than the electronic one and thus the muonic states are sensitive to the nuclear effects. Rigorously speaking, we note that the atomic linewidth (due to radiative transitions) scales as the mass of the orbiting particle m, while the nuclear contribution to the Lamb shift scales as m^3 and to the hyperfine structure as m^2. The width of the line is essentially responsible for the uncertainty in its measurement and an additional enhancement of the effect in respect to the width by the factor of $m_\mu/m_e \sim 207$ or $(m_\mu/m_e)^2 \sim 4 \cdot 10^4$ (depending on the effect under study) makes the muonic atomic spectroscopy a favorite tool to study the nuclear effects in atoms. The nuclear effects, especially for light nuclei, are reviewed in [55]. The target of the review is to tell atomic physicists what they can do for nuclear physics and what they can learn from it.

In contrast to the muonic atoms, the exotic atoms were presented in the former book [4] very well. Studies of exotic atoms, which contain at least one orbiting hadron, create an efficient atomic interface of particle physics and allows access for a very few quantities but with an incredibly high (for particle physics) accuracy. There were two main topics related to the exotic atoms reviewed in 2001: relativistic atoms (such as pionium, a bound system of $\pi^+\pi^-$) [56] and antiprotonic helium [57]. Studies of both pionium and antiprotonic helium were successfully developed.

During the last two years the DIRAC collaboration at CERN detected about 12 000 pionium events. The data have been partly analyzed and they lead to a result for the pionium lifetime with statistical uncertainty of 14% [58]. We note that in all previous experiments with a production of any relativistic atoms (see [56] for detail) they were produced in quantities not bigger than a few hundreds.

Researches on antiprotonic helium, a three-body atom consisitng of α-particle (nucleus) and an antiproton (in a highly excited circular state) and an electron, [57] have been also successfully continued. The ASACUSA collaboration measured hyperfine splitting [59] and tested the CPT symmetry by measuring in particular "antiprotonic Rydberg constants" which contains $m_{\bar{p}}$ instead of m_e and $e_{\bar{p}}$ instead of e_e. Two other collaborations working with antiprotons, ATHENA and ATRAP, are trying to create, trap and perform spectroscopic measurements of antihydrogen. They both recently succeeded in formation of so-called "cold" antihydrogen [61,62], i.e. an atom suited for spectroscopic studies. The untimate target of both the antihydrogen projects is the perform high resolution spectroscopic measurements on antihydrogen atoms and in particular to measure the $1s - 2s$ transition frequency. A comparison of the results with those for hydrogen will demonstrate whether the Rydberg constant is the same for atom and *anti*atom.

In our book the exotic atoms are presented in [63] (heavy pionic atoms) and in [53] (pionic, kaonic and antiprotonic hydrogen). These atoms were not considered in the previous book.

9   Quantum Mechanics of Hydrogen-Like Atoms: Tutorial

The clear structure of hydrogen and hydrogen-like atoms make them an important part of study of quantum mechanics. For this reason we include into the book two tutorial papers [64] and [65]. Quantum mechanics and quantum electrodynamics are perturbative theories and to work within their framework one needs first to solve an "unperturbed" problem which is essentially Coulomb problem. "To solve" means to learn the energy levels and the wave function of the states of interest. To construct a perturbative expansion one needs to find also a particular sum over all unperturbated states ζ which is called the Green function

$$G(E) = \sum_\zeta \frac{|\zeta\rangle \langle\zeta|}{E - E_\zeta} . \tag{6}$$

While the energy levels and the wave functions for the Coulomb problem are discussed in detail in a number of text books (see e.g. the classic book [1]) the Coulomb Green function is not considered there in detail. Lecture [64] is devoted to different presentations of the nonrelativistic and relativistic Coulomb Green function and its applications to several problems.

Sometimes it is possible to avoid any direct calculations, but still to obtain a result. An efficient tool for that is the virial theorem. As it is well known in a classical system bound by Newtonian gravitation, the average kinetic energy is one half of the potential energy and of opposite sign. Averaging over time is physically very close to a calculating an average over a quantum state and one can expect that a kind of virial theorem for the Coulomb interaction will be valid in nonrelativistic atomic physics. This approach may be actually developed even in the case of relativistic quantum mechanics and is discussed in lecture [65] in detail. Special attention is paid to applications.

10 About PSAS 2002 Conference

We Complete the introduction with a few words about our conference. The International conference on *Precision Physics of Simple Atomic Systems* (PSAS 2002) took place in St. Petersburg on June, 30–July, 3, 2002. The contributed papers formed the conference issue of the Canadian Journal of Physics **80**(11) (2002) and its table of contents can be found in Appendix to this book [6]. More detail about the meeting can be found in our website [66].

In completing a winter book for a summer conference, we note that some new results were achieved in between the conference (July, 2002) and submission of the book (December, 2002), such as the formation of antihydrogen, and we hope that they will be presented at *Hydrogen Atom, 3: Precision Physics of Simple Atomic systems* in 2004 and maybe will be considered in the next book.

Acknowledgements

Several papers of this book [38,41,53,55] present enlarged versions of contributions already published in the conference issue [67–70] and we are grateful to the Canadian Journal of Physics and to their publisher, the NRC Researche Press, who granted their permission for this publication. We also like to express our gratitude to the Springer-Verlag who agreed that part of the reviews would be published in the conference issue in a brief form. In particular we gratefully acknowledge that several of the figures used by our contributors in [38,41,53,55] have been already published by them in the Canadian Journal of Physics [67–70].

We are grateful to the Russian Foundation for Basic Research for support in organizing of the conference (under grant # 02-02-26086). We thank the members of the organizing committee and especially E.N. Borisov and V.A. Shelyuto for their great efforts in organization of the PSAS 2002 meeting. We are also grateful to another member of the Organizing Committee, Gordon Drake, for his crucial help in publishing of our conference issue. We would also like to thank our colleagues from the Russian Center of Laser Physics at the St. Petersburg State University, the D.I. Mendeleev Institute for Metrology (VNIIM) and the Max-Planck-Institut für Quantenoptik and especially O.S. Grynskii, A.V. Kurochkin, A.A. Man'shina and I.V. Schelkunov for their help.

References

1. H.A. Bethe and E.E. Salpeter: *Quantum Mechanics of One- and Two-electon Atoms* (Plenum, NY, 1977)
2. G.W. Series: *The Spectrum of Atomic Hydrogen: Advances* (World Sci., Singapore, 1988)
3. *The Hydrogen Atom*, Proceedings of the Simposium, Held in Pisa, Italy June, 30–July, 2, 1988. Edited by G. F. Bassani, M. Inguscio and T. W. Hänsch (Springer-Verlag, Berlin, Heidelberg, 1989), presented in CD'part of Ref. [4]
4. S.G. Karshenboim, F.S. Pavone, F. Bassani, M. Inguscio and T.W. Hänsch: *Hydrogen atom: Precision physics of simple atomic systems* (Springer, Berlin, Heidelberg, 2001)
5. S.G. Karshenboim: In *Atomic Physics* **17** (AIP conference proceedings 551) Ed. by E. Arimondo et al. (AIP, 2001), p. 238
6. *This book*, pp. 211–211
7. F. Biraben, T.W. Hänsch, M. Fischer, M. Niering, R. Holzwarth, J. Reichert, Th. Udem, M. Weitz, B. de Beauvoir, C. Schwob, L. Jozefowski, L. Hilico, F. Nez, L. Julien, O. Acef, J.-J. Zondy, and A. Clairon: In Ref. [4], p. 17
8. G. Drake: In Ref. [4], p. 57
9. S.G. Karshenboim: In Ref. [4], p. 335
10. V.A. Yerokhin and V. M. Shabaev: Phys. Rev. A **64**, 012506 (2001)
11. S.G. Karshenboim and V.G. Ivanov: Phys. Lett. B **524**, 259 (2002); Euro. Phys. J. D **19**, 13 (2002)
12. M. Fischer, N. Kolachevsky, S.G. Karshenboim and T.W. Hänsch: Can. J. Phys. **80**, 1225 (2002)
13. N. Kolachevsky, M. Fischer, S.G. Karshenboim and T.W. Hänsch: in preparation
14. N.E. Rothery and E.A. Hessels: Phys. Rev. A **61**, 044501 (2000)
15. S.G. Karshenboim: In *This book*, pp. 141–162
16. K. Pachucki and J. Sapirstein: J. Phys. B: At. Mol. Opt. Phys. **35**, 1783 (2002)
17. G.W.F. Drake: Can. J. Phys. **80**, 1195 (2002)
18. M.C. George, L. D. Lombardi, and E.A. Hessels: Phys. Rev. Lett. **87**, 173002 (2001)
19. F. Minardi, G. Bianchini, P. Cancio Pastor, G. Giusfredi, F.S. Pavone, and M. Inguscio, Phys. Rev. Lett. **82**, 1112 (1999);
    P.C. Pastor, P. De Natale, G. Giusfredi, F.S. Pavone, and M. Inguscio: In Ref. [4], p. 314
20. J. Castillega, D. Livingston, A. Sanders, and D. Shiner, Phys. Rev. Lett. **84**, 4321 (2000)
21. J. Wen, Ph.D. thesis, Harvard University, 1996 (unpublished)
22. W. Frieze, E.A. Hinds, V.W. Hughes, and F.M.J. Pichanick, Phys. Rev. A **24**, 279 (1981)
23. K.-P. Jungmann: In Ref. [4], p. 81
24. R.S. Conti, R.S. Vallery, D.W. Gidley, J.J. Engbrecht, M. Skalsey, and P.W. Zitzewitz: In Ref. [4], p. 103
25. P. Crivelli: Can. J. Phys. **80**, p. 1281 (2002);
    M. Chiba, J. Nakagawa, H. Tsugawa, R. Ogata, and T. Nishimura: Can. J. Phys. **80**, p. 1287 (2002)
26. S.G. Karshenboim and V.G. Ivanov: Can. J. Phys. **80**, 1305 (2002)
27. S.I. Eidelman, S. G. Karshenboim and V A. Shelyuto: Can. J. Phys. **80**, 1297 (2002)
28. E.G. Myers: In Ref. [4], p. 179

29. G. Werth, H. Häffner, N. Hermanspahn, H.-J. Kluge, W. Quint, and J. Verdú: In Ref. [4], p. 204

30. T. Stöhlker, D. Banaś, H. Beyer and A. Gumberidze: In *This book*, pp. 115–137

31. S.G. Karshenboim: In Ref. [2], p. 651

32. T. Beier, H. Häffner, N. Hermanspahn, S.G. Karshenboim, H.-J. Kluge, W. Quint, S. Stahl, J. Verdú, and G. Werth: Phys. Rev. Lett. **88**, 011603 (2002)

33. J.L. Verdú, S. Djekic, T. Valenzuela, H. Häffner, W. Quint, H.J. Kluge, and G. Werth: Can. J. Phys. **80**, 1233 (2002)

34. V.A. Yerokhin, P. Indelicato, and V.M. Shabaev: Can. J. Phys. **80**, p. 1249 (2002)

35. S.G. Karshenboim and A.I. Milstein: Phys. Lett. B **549**, 321 (2002)

36. M. Redshaw and E.G. Myers: Phys. Rev. Lett. **88**, 023002 (2002)

37. A. Wicht, J.M. Hensley, E. Sarajlic, and S. Chu: In *Proceedings of the 6th Symposium Frequency Standards and Metrology*, ed. by P. Gill (World Sci., 2002) p. 193

38. S. Rainville, J.K. Thompson, and D.E. Pritchard: In *This book*, pp. 177–197

39. P. Mohr and B.N. Taylor: In Ref. [4], p. 145

40. T. Kinoshita: In Ref. [4], p. 157

41. S. Redin, G.W. Bennett, B. Bousquet, H.N. Brown, G. Bunce, R.M. Carey, P. Cushman, G.T. Danby, P.T. Debevec, M. Deile, H. Deng, W. Deninger, S.K. Dhawan, V.P. Druzhinin, L. Duong, E. Efstathiadis, F.J.M. Farley, G.V. Fedotovich, S. Giron, F. Gray, D. Grigoriev, M. Grosse-Perdekamp, A. Grossmann, M.F. Hare, D W. Hertzog, X. Huang, V.W. Hughes, M. Iwasaki, K. Jungmann, D. Kawall, M. Kawamura, B.I. Khazin, J. Kindem, F. Krinen, I. Kronkvist, A. Lam, R. Larsen, Y.Y. Lee, I. B. Logashenko, R. McNabb, W. Meng, J. Mi, J.P. Miller, W.M. Morse, D. Nikas, C.J.G. Onderwater, Yu.F. Orlov, C. Ozben, J. Paley, Q. Peng, J. Pretz, R. Prigl, G. zu Putlitz, T. Qian, O. Rind, B.L. Roberts, N M. Ryskulov, P. Shagin, S. Sedykh, Y.K. Semertzidis, Yu.M. Shatunov, E.P. Solodov, E.P. Sichtermann, M. Sossong, A. Steinmetz, L.R. Sulak, C. Timmermans, A. Trofimov, D. Urner, P. von Walter, D. Warburton, D. Winn, A. Yamamoto, and D. Zimmerman: In *This book*, pp. 163–174

42. M. Davier, S. Eidelman, A. Höcker and Z.Q. Zhang: Eur. Phys. J. C**27**, 497 (2003). See also a short discussion in [27]

43. T. Kinoshita and M. Nio: Phys. Rev. Lett. **90**, 021803 (2003)

44. V.A. Dzuba, V.V. Flambaum, M.T. Murphy, and J.K. Webb: In Ref. [4], p. 564

45. D.A. Varshalovich, A.V. Ivanchik, A.V. Orlov, A.Y. Potekhin and P. Petitjean: In *This book*, pp. 199–209

46. J.K. Webb, M.T. Murphy, V.V. Flambaum, S.J. Curran: Astrophys. J. Supp. **283**, 565 (2003)

47. T. Udem, J. Reichert, R. Holzwarth, S. Diddams, D. Jones, J. Ye, S. Cundiff, T. Hänsch, and J. Hall: In Ref. [4], p. 125.

48. T. Udem, S.A. Diddams, K.R. Vogel, C.W. Oates, E.A. Curtis, W.D. Lee, W.M. Itano, R.E. Drullinger, J.C. Bergquist, and L. Hollberg, Phys. Rev. Lett. **86**, 4996 (2001);
S. Bize, S.A. Diddams, U. Tanaka, C.E. Tanner, W.H. Oskay, R.E. Drullinger, T.E. Parker, T.P. Heavner, S.R. Jefferts, L. Hollberg, W.M. Itano, D.J. Wineland, and J.C. Bergquist: Phys. Rev. Lett. **90**, 150802 (2003)

49. J. Stenger, C. Tamm, N. Haverkamp, S. Weyers, and H.R. Telle, Opt. Lett. **26**, 1589 (2001).

50. G. Wilpers, T. Binnewies, C. Degenhardt, U. Sterr, J. Helmcke, and F. Riehle, Phys. Rev. Lett. **89**, 230801 (2002)

51. H. Marion, F. Pereira Dos Santos, M. Abgrall, S. Zhang, Y. Sortais, S. Bize, I. Maksimovic, D. Calonico, J. Grünert, C. Mandache, P. Lemonde, G. Santarelli, Ph. Laurent, A. Clairon, and C. Salomon: Phys. Rev. Lett. **90**, 150801 (2003)
52. S G. Karshenboim: Can. J. Phys. **78**, 639 (2000).
53. T.S. Jensen and V.E. Markushin: In *This book*, pp. 37–57.
54. S. Geer: Phys. Rev. D **57**, 6989 (1998); D **59**, 039903 (E) (1998); B. Autin, A. Blondel and J. Ellis (eds.): *Prospective study of muon storage rings at CERN*, Report CERN-99-02.
55. J.L. Friar: In *This book*, pp. 58–79.
56. L. Nemenov: In Ref. [4], p. 223.
57. T. Yamazaki: In Ref. [4], p. 246
58. L. Nemenov: private communication.
59. E. Widmann, J. Eades, T. Ishikawa, J. Sakaguchi, T. Tasaki, H. Yamaguchi, R.S. Hayano, M. Hori, H.A. Torii, B. Juhasz, D. Horvath, and T. Yamazaki: Phys. Rev. Lett. **89**, 243402 (2002)
60. M. Hori, J. Eades, R. S. Hayano, T. Ishikawa, J. Sakaguchi, E. Widmann, H. Yamaguchi, H.A. Torii, B. Juhasz, D. Horvath, and T. Yamazaki: Phys. Rev. Lett. **87**, 093401 (2001)
61. M. Amoretti, C. Amsler, G. Bonomi, A. Bouchta, P. Bowe, C. Carraro, C.L. Cesar, M. Charlton, M.J.T. Collier, M. Doser, V. Filippini, K.S. Fine, A. Fontana, M.C. Fujiwara, R. Funakoshi, P. Genova, J.S. Hangst, R.S. Hayano, M.H. Holzscheiter, L.V. Jorgensen, V. Lagomarsino, R. Landua, D. Lindelöf, E. Lodi Rizzini, M. Macri, N. Madsen, G. Manuzio, M. Marchesotti, P. Montagna, H. Pruys, C. Regenfus, P. Riedler, J. Rochet, A. Rotondi, G. Rouleau, G. Testera, A. Variola, T.L. Watson and D.P. van der Werf: Nature **419**, 456 (2002)
62. G. Gabrielse, N.S. Bowden, P. Oxley, A. Speck, C.H. Storry, J.N. Tan, M. Wessels, D. Grzonka, W. Oelert, G. Schepers, T. Sefzick, J. Walz, H. Pittner, T.W. Hänsch, and E.A. Hessels: Phys. Rev. Lett. **89**, 213401 (2002); Phys. Rev. Lett. **89**, 233401 (2002).
63. T. Yamazaki: In *This book*, pp. 80–93.
64. L.N. Labzowsky and D.A. Solovyev: In *This book*, pp. 15–34.
65. V M. Shabaev: In *This book*, pp. 97–113.
66. http://psas2002.vniim.ru or http://home.rclph.spbu.ru/psas2002.
67. S. Rainville, J.K. Thompson, D.E. Pritchard: Can. J. Phys. **80**, 1329 (2002).
68. S. Redin, R.M. Carey, E. Efstathiadis, M.F. Hare, X. Huang, F. Krinen, A. Lam, J.P. Miller, J. Paley, Q. Peng, O. Rind, B.L. Roberts, L.R. Sulak, A. Trofimov, G.W. Bennett, H.N. Brown, G. Bunce, G.T. Danby, R. Larsen, Y.Y. Lee, W. Meng, J. Mi, W.M. Morse, D. Nikas, C. Ozben, R. Prigl, Y.K. Semertzidis, D. Warburton, V.P. Druzhinin, G.V. Fedotovich, D. Grigoriev, B.I. Khazin, I.B. Logashenko, N.M. Ryskulov, Yu.M. Shatunov, E.P. Solodov, Yu.F. Orlov, D. Winn, A. Grossmann, K. Jungmann, G. zu Putlitz, P. von Walter, P.T. Debevec, W. Deninger, F. Gray, D.W. Hertzog, C.J.G. Onderwater, C. Polly, S. Sedykh, M. Sossong, D. Urner, A. Yamamoto, B. Bousquet, P. Cushman, L. Duong, S. Giron, J. Kindem, I. Kronkvist, R. McNabb, T. Qian, P. Shagin, C. Timmermans, D. Zimmerman, M. Iwasaki, M. Kawamura, M. Deile, H. Deng, S.K. Dhawan, F.J.M. Farley, M. Grosse-Perdekamp, V.W. Hughes, D. Kawall, J. Pretz, E.P. Sichtermann, and A. Steinmetz: Can. J. Phys. **80**, 1355 (2002).
69. V.E. Markushin: Can. J. Phys. **80**, 1271 (2002).
70. J.L. Friar: Can. J. Phys. **80**, 1337 (2002).

Part I

The Hydrogen Atom

Coulomb Green Function and Its Applications in Atomic Theory

L.N. Labzowsky[1,2] and D.A. Solovyev[1]

[1] St. Petersburg State University 198904, Petrodvoretz, St. Petersburg, Russia
[2] Petersburg Nuclear Physics Institute 188350, Gatchina, St. Petersburg, Russia

Abstract. The applications of the Coulomb Green function (CGF) to the calculation of different atomic properties are reviewed. The different representations for the Coulomb Green function including Sturmian expansions are considered.

1 Coulomb Green Function for the Schrödinger Equation

The Coulomb Green function is a convenient tool for the evaluation of sums over the entire spectrum of the Schrödinger equation. These sums arise usually when perturbation theory is applied. The Green function approach helps to express these sums in a closed explicit analytic form, what is very useful for analysis and for the validation of the numerical calculations.

We consider the nonrelativistic Coulomb problem, i.e. the stationary Schrödinger equation

$$\widehat{H}\psi \equiv -(\Delta + \frac{Z}{r})\psi = E\psi \,, \tag{1}$$

where atomic units are used. The well-known solutions of this equation can be presented in spherical coordinates $\boldsymbol{r} \equiv r, \Omega(\Omega \equiv \theta\varphi)$ in the form

$$\psi(\boldsymbol{r}) = R_{nl}(r)Y_{lm}(\Omega) \,, \tag{2}$$

where Y_{lm} are spherical functions, and l, m the angular quantum numbers. In the case of the discrete spectrum n is the principal quantum number $n = 1, 2, ...; n \geq l + 1$. The eigenvalues corresponding to the solutions (2) are

$$E_n = -\frac{Z^2}{2n^2} \,. \tag{3}$$

For the solution of many problems in atomic theory the application of the Green function, corresponding to (1) appears to be useful. The Green function is called Coulomb Green function (CGF) and is the solution of the equation

$$(\widehat{H} - E)G_E(\boldsymbol{r}; \boldsymbol{r'}) = \delta(\boldsymbol{r} - \boldsymbol{r'}) \,. \tag{4}$$

The CGF always can be written as a spectral expansion

$$G_E(\boldsymbol{r}; \boldsymbol{r'}) = \sum_i \frac{\psi_i^*(\boldsymbol{r})\psi_i(\boldsymbol{r'})}{E_i - E} \,, \tag{5}$$

where the sum is extended over the total spectrum of the Hamiltonian (1), including the continuous spectrum. The formula (5) can be easily verified with the use of the completeness of the system of the eigenfunctions of the Hamiltonian (1):

$$\sum_i \psi_i^*(\boldsymbol{r})\psi_i(\boldsymbol{r}') = \delta(\boldsymbol{r} - \boldsymbol{r}') \ . \tag{6}$$

The expression (5) is not well suited for practical purposes. For most of the applications it is enough to have the closed expressions for the radial part of the CGF defined by the partial wave expansion

$$G_E(\boldsymbol{r};\boldsymbol{r}') = \sum_{lm} \frac{1}{rr'} r') G_{El}(r;r') Y_{lm}^*(\Omega) Y_{lm}(\Omega') \ . \tag{7}$$

We will also use the partial wave expansion for the Dirac δ-function;

$$\delta(\boldsymbol{r} - \boldsymbol{r}') = \frac{1}{rr'} \delta(r - r') \sum_{lm} Y_{lm}^*(\Omega) Y_{lm}(\Omega') \tag{8}$$

Inserting the expansions (7) and (8) in (4) and separating the angular variables we obtain

$$\left[-\frac{1}{2}\frac{d^2}{dr^2} + \frac{l(l+1)}{2r^2} - \frac{Z}{r} - E \right] G_{El}(r;r') = \delta(r - r') \ . \tag{9}$$

The standard approach to the construction of the Green function for an arbitrary linear second-order differential equation is the following. Consider the equation

$$\hat{L}(x)y(x) \equiv \left[\frac{d}{dx}\left(p(x)\frac{d}{dx}\right) + q(x) - \lambda \right] y(x) = 0 \tag{10}$$

with the homogeneous boundary equations at the ends of the interval x_1, x_2:

$$\alpha_1 y(x_1) + \alpha_2 y'(x_1) = 0 \ , \tag{11}$$

$$\beta_1 y(x_2) + \beta_2 y'(x_2) = 60 \ . \tag{12}$$

The corresponding equation for the Green function looks like

$$\hat{L}(x)g(x,x') = \delta(x - x') \ . \tag{13}$$

The solution of (13) is

$$g(x,x') = y_1(x_<)y_2(x_>) \ , \tag{14}$$

where $y_1(x)$, $y_2(x)$ are two linearly-independent solutions of (10). The solution $y_1(x)$ satisfies the boundary condition (11) and the solution $y_2(x)$ satisfies the boundary condition (12). The notation $x_<, x_>$ corresponds to the smaller or larger of the arguments x_1, x_2. The Green function (14) is normalized by the requirement that it should satisfy (13) at $x = x'$ as well. This requirement leads to the condition

$$W(x) \equiv y_1(x)y_2'(x) - y_2(x)y_1'(x) = \frac{1}{p(x)} . \tag{15}$$

The left-hand side of Eq. (15) presents the Wronski determinant (Wronskian) of (10).

Let us now return to (9). By the substitution $r = \frac{x}{2\sqrt{-2E}}$ the left-hand side of this equation can be reduced to the Whittaker equation

$$\frac{d^2U(x)}{dx^2} + \left[\frac{\nu}{x} + \frac{1/4 - \mu^2}{x^2} - \frac{1}{4}\right]U(x) = 0 , \tag{16}$$

where $\nu \equiv \frac{Z}{\sqrt{-2E}}$, $\mu = l + \frac{1}{2}$. Equation (16) is solved on the interval $0 \le x \le \infty$ and the boundary conditions reduce to the requirements that $U(x)$ should be finite at $x = 0$ and should turn to zero at $x = \infty$. The linearly-independent solutions that satisfy these requirements are the Whittaker functions $M_{\nu,\mu}(x)$ and $W_{\nu,\mu}(x)$ correspondingly. These functions can be expressed through the degenerate hypergeometric function:

$$M_{\nu,\mu}(x) = x^{\mu+\frac{1}{2}}e^{-\frac{x}{2}}F\left(\mu - \nu + \frac{1}{2}, 2\mu + 1; x\right) , \tag{17}$$

$$W_{\nu,\mu}(x) = \frac{\Gamma(-2\mu)}{\Gamma(\frac{1}{2} - \mu - \nu)}M_{\nu,\mu}(x) + \frac{\Gamma(2\mu)}{\Gamma(\frac{1}{2} + \mu - \nu)}M_{-\nu,-\mu}(x) . \tag{18}$$

Asymptotic expressions for the Whittaker functions by $x \to \infty$ are

$$M_{\nu,\mu}(x) \to x^{-\nu}e^{\frac{x}{2}} , \tag{19}$$

$$W_{\nu,\mu}(x) \to x^{\nu}e^{-\frac{x}{2}} . \tag{20}$$

It follows from Eqs. (19), (20) that the function $W_{\nu,\mu}$ satisfies the boundary conditions at $x = \infty$.

Now we evaluate the Wronskian W in (15). For (10), we have $p(x) = -\frac{1}{2}$, i.e. the Wronskian does not depend on x and we can evaluate it at arbitrary values of x. It is most convenient to choose $x = 0$. Then, setting $y_1 = M_{\nu,\mu}$ and $y_2 = W_{\nu,\mu}$ and evaluating the derivatives in Eq. (15) by $x \to 0$, one obtains

$$M_{\nu,\mu}(2Zr/\nu)\frac{d}{dr}W_{\nu,\mu}(2Zr/\nu) - W_{\nu,\mu}(2Zr/\nu)\frac{d}{dr}M_{\nu,\mu}(2Zr/\nu)$$
$$= -\frac{2Z}{\nu}\frac{\Gamma(2\mu+1)}{\Gamma(\frac{1}{2}+\mu-\nu)} . \tag{21}$$

Thus the normalization factor for the Green function is $\frac{\nu}{Z}\frac{\Gamma(\frac{1}{2}+\mu-\nu)}{\Gamma(2\mu+1)}$ and finally the normalized expression for $G_{El}(r, r')$ looks like [1]

$$G_{El}(r, r') = \frac{\nu}{Z}\frac{\Gamma(l+1-\nu)}{\Gamma(2l+2)}M_{\nu,l+\frac{1}{2}}(2Zr_</\nu)W_{\nu,l+\frac{1}{2}}(2Zr_>/\nu) . \tag{22}$$

This expression has the necessary poles at integer values of $\nu = n$, as expected. The condition $\nu = n$ leads again to the Balmer's formula (3).

The closed expression (22) still is not very convenient for calculations due to the presence of the complicated arguments $r_<, r_>$. Therefore, in applications the following integral representation is often used [1]

$$
M_{\nu,l+\frac{1}{2}}(at)W_{\nu,l+\frac{1}{2}}(at) = \frac{t(ab)^{\frac{1}{2}}\Gamma(2l+2)}{\Gamma(l+1-\nu)}
$$

$$
\times \int_0^\infty ds(\operatorname{cth}\frac{s}{2})^{2\nu}\exp\left[-\frac{1}{2}(a+b)t\cdot\operatorname{ch}(s)\right] I_{2l+1}(t(ab)^{1/2}\operatorname{sh}(s)) , \qquad (23)
$$

where I_{2l+1} is the modified Bessel's function. The substitution $\operatorname{ch}(s) = \xi$ leads to another convenient representation [1],

$$
M_{\nu,l+\frac{1}{2}}(at)W_{\nu,l+\frac{1}{2}}(bt) = \frac{t(ab)^{\frac{1}{2}}\Gamma(2l+2)}{2\Gamma(l+1-\nu)} \times
$$

$$
\int_1^\infty d\xi(1+\xi)^{\nu-\frac{1}{2}}(1-\xi)^{-\nu-\frac{1}{2}}\exp\left[-\frac{1}{2}(a+b)+\xi\right] I_{2l+1}(t(ab)^{1/2}(\xi^2-1)^{1/2}) .
$$

$$
\qquad (24)
$$

Expanding I_{2l+1} at $\xi = 0$ one can easily see that the integral in Eq. (24) is convergent under the requirement

$$
\operatorname{Re}(l+1-\nu) > 0 . \qquad (25)
$$

2   Sturmian Expansion

In some cases it is more convenient to consider the radial Schrödinger equation as a generalized Sturm-Liouville problem [2],

$$
-\frac{1}{2r^2}\left(r^2\frac{d\Phi(r)}{dr}\right) + \frac{l(l+1)}{2r^2}\Phi(r) - E\Phi(r) = \frac{Z}{r}\Phi(r) . \qquad (26)
$$

In (26) the energy parameter is fixed and one looks for the allowed values of the parameter Z which corresponds to the solutions, satisfying the boundary conditions. These solutions Φ_{Zl} or the eigenfunctions of the generalized Sturm-Liouville problem, called also Sturmian functions, present a complete system of functions which can be used for the expansion of an arbitrary function. They are orthogonal with the weight $1/r$,

$$
\int_0^\infty \Phi_{Z_1l}(r)\Phi_{Z_2l}(r)rdr = \delta_{Z_1 Z_2} . \qquad (27)
$$

Using the substitutions $\Phi_{Zl}(r) = \frac{1}{r}U_{Zl}(r)$ and $r = \frac{\rho}{2\sqrt{-2E}}$ we arrive again at the Whittaker equation (16). The general solution of the Whittaker equation is

$$
U_{Zl} = \frac{c_1}{\rho}M_{\nu,\mu}(\rho) + \frac{c_2}{\rho}W_{\nu,\mu}(\rho) . \qquad (28)
$$

The solution should be finite at $r \to 0$; therefore $c_2 = 0$. From the same requirement at $r \to \infty$ it follows

$$\mu - \nu + \frac{1}{2} = -n_r , \tag{29}$$

where n_r is the radial quantum number $n_r = 0, 1, \ldots$. Equation (29) corresponds to the condition that the expansion of the hypergeometric function F in Eq. (17) contains a limited number of the terms. From (29) we have

$$Z \equiv Z_{nl} = (n_r + l + 1)\sqrt{-2E} . \tag{30}$$

This means that for the bound states ($E < 0$) the generalized Sturm-Liouville equation possess only the discrete spectrum of Z values that are defined by (30). The radial quantum number n_r is connected with the principal quantum number n by the relation

$$n = n_r + l + 1 . \tag{31}$$

The Sturmian functions, orthonormalized according to (27) look like

$$\Phi_{n_r l}(r) = \frac{1}{(2l+1)!} \left[\frac{\Gamma(2l + n_r + 2)}{\Gamma(n_r + 1)} \right]^{\frac{1}{2}} \frac{1}{r} M_{n_r + l + 1, l + \frac{1}{2}} \left(\frac{2\sqrt{-2E}}{Z} r \right) . \tag{32}$$

The completeness of the functions $\Phi_{n_r l}(r)$ means

$$\sum_{n_r=0}^{\infty} \Phi_{n_r l}(r) \Phi_{n_r l}(r') = \frac{1}{r} \delta(r - r') = \frac{1}{r'} \delta(r - r') . \tag{33}$$

The solutions of (26) with the different allowed Z values and the fixed value of E do not describe any real atomic states. However they present a convenient basis for expansions since the continuous part of the spectrum is absent. In particular, the Sturmian basis appears to be useful for the expansion of the radial part of the Coulomb Green function [3]. This expansion can be written in the form

$$\frac{1}{rr'} G_{El}(r; r') = \sum_{n_r=0}^{\infty} \frac{\Phi_{n_r l}(r) \Phi_{n_r l}(r')}{Z_{n_r l} - Z} \tag{34}$$

To verify (34) one can act on the both sides of this equation with the operator

$$\widehat{H}'(r, E) - \frac{Z}{r} ,$$

where $\widehat{H}'(r, E)$ is the operator from the left-hand side of (26). Then, using the completeness condition (33) we arrive at the equation

$$\left(\widehat{H}'(r, E) - \frac{Z}{r} \right) \frac{1}{rr'} G_{El}(r; r') = \frac{1}{rr'} \delta(r - r') , \qquad (35)$$

which coincides with (9).

The Sturmian expansion of the Coulomb Green function can be presented also in another form by introducing the functions

$$R_{nl}\left(\frac{2r}{\nu} \right) = \frac{1}{r} \sqrt{\frac{Z}{\nu n}} \frac{1}{(2l+1)!} \sqrt{\frac{\Gamma(n+l+1)}{\Gamma(n-l)}} M_{n,l+\frac{1}{2}}\left(\frac{2r}{\nu} \right) . \qquad (36)$$

At $\nu = n$ these functions coincide with the normalized radial hydrogenic wave functions. Comparison of (36) with (32) yields:

$$\Phi_{n_r l}(r) = \sqrt{\frac{\nu n}{Z}} R_{nl}\left(\frac{2r}{\nu} \right) . \qquad (37)$$

The expansion (34) can be rewritten in the form

$$\frac{1}{rr'} G_{El}(r; r') = \frac{\nu^2}{Z^2} \sum_{n=l+1}^{\infty} \frac{n}{n-\nu} R_{nl}\left(\frac{2r}{\nu} \right) R_{nl}\left(\frac{2r}{\nu} \right) . \qquad (38)$$

Here $\nu = \frac{Z}{\sqrt{-2E}}$ and the poles of the Green function corresponding to the hydrogenic energy levels are seen explicitly. The expansion (38) is especially convenient for obtaining the modified Green function, the radial part of which is defined as

$$\tilde{G}_{E_n l}(r; r') = \lim_{E \to E_n} \left[G_{El}(r; r') - rr' \frac{R_{nl}(r)R_{nl}(r)}{E_n - E} \right] . \qquad (39)$$

Setting $\nu = n - \epsilon$ in Eq. (38) and performing an expansion in ϵ, we obtain

$$\tilde{G}_{E_n l}(r; r') = rr' \frac{n^2}{Z^2} \sum_{n'=l+1 \, (n' \neq n)} \left\{ \frac{n'}{n'-n} R_{n'l}\left(\frac{2r}{n} \right) R_{n'l}\left(\frac{2r'}{n} \right) \right.$$
$$+ \frac{1}{2} rr' \frac{n^2}{Z^2} R_{n'l}\left(\frac{2r}{n} \right) R_{n'l}\left(\frac{2r'}{n} \right)$$
$$\left. - 2rr' \frac{n}{Z^2} \left[r R'_{n'l}\left(\frac{2r}{n} \right) R_{n'l}\left(\frac{2r'}{n} \right) + r' R_{n'l}\left(\frac{2r}{n} \right) R'_{n'l}\left(\frac{2r'}{n} \right) \right] \right\} , \qquad (40)$$

where $R'_{n'l}(x) \equiv \frac{d}{dx} R_{nl}(x)$.

The modified Green function corresponds to the definite atomic bound state $(E = E_n)$. Such functions are applied widely.

3   Two-Photon Decay of Atomic Levels

As a first application of the Green function method we consider the two-photon decay of excited hydrogen atom levels [4]. The probability of the two-photon decay integrated over the emitted photon directions and summed over the polarizations looks like (in relativistic units)

$$dw_{AA'}^{(2)} = \frac{8\omega^3(\omega_{AA'} - \omega)^3}{9\pi}\alpha^2 \sum_{ik} |(U_{ik})_{A'A}|^2 \, d\omega \, . \tag{41}$$

Here the symbols A, A' denote the initial and final atomic states, $\omega_{AA'}$ is the energy difference between these states $\omega_{AA'} = E_A - E_{A'}$, ω and $\omega' = \omega_{AA'} - \omega$ are the frequencies of the two emitted photons and α is the fine structure constant. The matrix element of the tensor $(U_{ik})_{A'A}$ is defined as

$$(U_{ik})_{A'A} = \sum_s \left[\frac{(x_i)_{A's}(x_k)_{sA}}{E_s - E_A + \omega} + \frac{(x_k)_{A's}(x_i)_{sA}}{E_s - E_{A'} - \omega} \right] , \tag{42}$$

where the summation is extended over the total spectrum of $\widehat{H}$. Rewriting Eq. (42) in the spherical components we obtain

$$\sum_{qq'} |(U_{qq'})_{A'A}|^2 = \sum_{qq'} (-1)^{q+q'} |(U_{q'q})_{A'A}|^2 \, . \tag{43}$$

$$(U_{qq'})_{A'A} \equiv \sum_s \left[\frac{(r_{q'})_{A's}(r_q)_{sA}}{E_s - E_A + \omega} + \frac{(r_q)_{A's}(r_{q'})_{sA}}{E_s - E_A + \omega'} \right] , \tag{44}$$

where r_q are the spherical components of the radius-vector $\boldsymbol{r}$.

Remembering now the spherical expansion (5) of the Coulomb Green function we can present (44) in the form

$$(U_{q'q})_{A'A} = \left(r'_q G_{E_A - \omega}(\boldsymbol{r}'; \boldsymbol{r}) r_q \right)_{A'A} + \left(r_q G_{E_A - \omega'}(\boldsymbol{r}'; \boldsymbol{r}) r_{q'} \right)_{A'A} \tag{45}$$

In (45) we insert the partial wave expansion for the Green function (7), the expression for the atomic wave functions in the standard form (2) and the spherical radius-vector components

$$r_q = \sqrt{\frac{4\pi}{3}} r Y_{1q}(\Omega) \, . \tag{46}$$

Then the integration over all angles yields:

$$(U_{q'q})_{n'l'm',nlm} =$$

$$(-1)^{m'} \sqrt{(2l+1)(2l'+1)} \sum_{\lambda\mu} (2\lambda+1) \begin{pmatrix} l'1\lambda \\ 000 \end{pmatrix} \begin{pmatrix} \lambda 1l \\ 000 \end{pmatrix}$$

$$\times \left[S_{n'l',nl}^\lambda(\omega) \begin{pmatrix} l'1\lambda \\ \overline{m}'q'\mu \end{pmatrix} \begin{pmatrix} \lambda 1l \\ \overline{\mu}qm \end{pmatrix} + S_{n'l',nl}^\lambda(\omega') \begin{pmatrix} l'1\lambda \\ \overline{m}'q\mu \end{pmatrix} \begin{pmatrix} \lambda 1l \\ \overline{\mu}q'm \end{pmatrix} \right] , \tag{47}$$

where

$$S_{n'l',nl}^\lambda(\omega) = \int_0^\infty r^2 dr \int_0^\infty r'^2 dr' R_{n'l'}(r') G_{E_{nl} - \omega, \lambda}(r'; r) R_{nl}(r) \, . \tag{48}$$

Inserting (47) in (42) and performing the summation over the indices m' and the averaging over the indices m we arrive at the expression for the differential two-photon transition probability

$$
dw^{(2)}_{n'l',nl} = \frac{1}{2l+1}\sum_{mm'} dw^{(2)}_{n'l'm',nlm} = \frac{8\omega^3\omega'^3}{9\pi}\alpha^6(2l'+1)
$$

$$
\times \sum_{\lambda\lambda a}(2l+1)(2\lambda'+1)(2a+1)^2 \begin{pmatrix} l'1\lambda \\ 000 \end{pmatrix}\begin{pmatrix} \lambda 1 l \\ 000 \end{pmatrix}\begin{pmatrix} l'1\lambda' \\ 000 \end{pmatrix}\begin{pmatrix} \lambda'1 l \\ 000 \end{pmatrix}
$$

$$
\times \Bigg(\Big[S^{\lambda}_{n'l',nl}(\omega)S^{\lambda'}_{n'l',nl}(\omega) + S^{\lambda}_{n'l',nl}(\omega')S^{\lambda'}_{n'l',nl}(\omega') \Big]\begin{Bmatrix} 11a \\ l'l\lambda \end{Bmatrix}\begin{Bmatrix} 11a \\ l'l\lambda' \end{Bmatrix}
$$

$$
+ \Big[S^{\lambda}_{n'l',nl}(\omega)S^{\lambda'}_{n'l',nl}(\omega') + S^{\lambda}_{n'l',nl}(\omega')S^{\lambda'}_{n'l',nl}(\omega) \Big]\begin{Bmatrix} 11a \\ ll'\lambda' \end{Bmatrix}\begin{Bmatrix} 11a \\ l'l\lambda \end{Bmatrix} \Bigg) d\omega . \quad (49)
$$

For the two-photon decay of the $2s$ level $n'l' = 10$, $nl = 20$ and Eq. (49) reduces to

$$
dw^{(2)}_{10,20} = \frac{8(\omega\omega')^3}{27\pi}\alpha^6 \left[S^1_{10,20}(\omega) + S^1_{10,20}(\omega') \right] . \quad (50)
$$

It is convenient in this case to employ the representation (23) for the radial part of the Coulomb Green function. Then

$$
S^1_{10,20}(\omega) = 2\sqrt{2}\left[Q^1_{00}\left(\nu,\frac{\nu}{2};\nu\right) - \frac{1}{2}Q^1_{01}\left(\nu,\frac{\nu}{2};\nu\right) \right] , \quad (51)
$$

where

$$
Q^l_{s's}(\beta,\beta';\nu) = \int_0^\infty\int_0^\infty\int_0^\infty dr'drdx(r')^{s'+\frac{7}{2}}(r)^{s+\frac{7}{2}} \times
$$

$$
\times\exp\left\{-\frac{1}{\nu}\beta'r' + \beta r + (r+r')chx\right\}\left(cth\frac{x}{2}\right)^{2\nu}I_{2l+1}\left(\frac{2\sqrt{rr'}}{\nu}sh(x)\right),
$$
$$
\beta = \frac{\nu}{n}, \beta' = \frac{\nu}{n'}, \nu = (-2(E_{nl} - \omega))^{-\frac{1}{2}} . \quad (52)
$$

A similar representation should be used for $S^1_{10,20}(\omega')$.

The integrals $Q^l_{s's}$ can be evaluated analytically by expanding the Bessel's function I_{2l+1}; the integral over x should be evaluated in the end. It is convenient to start with the evaluation of the integral

$$
Q^l_{l-1,l-1}(\beta,\beta';\nu) = \frac{2^{2l+1}(2l+1)!(\nu)^{2l+3}}{(l+1-\nu)[(1+\beta)(1+\beta')]^{2l+2}}
$$

$$
\times {}_2F_1\left(2l+2,l+1-\nu;l+2-\nu;\frac{(\beta-1)(\beta'-1)}{(\beta+1)(\beta'+1)}\right) . \quad (53)
$$

All other explicit expressions for integrals $Q^l_{s',s}$ can be obtained by derivating the latter formula.

The total probability of the two-photon decay equals to

$$w_{AA'}^{(2)} = \int_0^{\omega_{AA'}} dw_{AA'}^{(2)}(\omega) \, . \tag{54}$$

The order of magnitude of the two-photon $2E1$ transition $2s \to 1s$ is $m\alpha^2(\alpha Z)^6$ r.u. $= (\alpha Z)^6$ a.u.

4 Lamb Shift in the Hydrogen Atom

The standard way of evaluating the Lamb shift (LS) in the hydrogen atom and in low Z hydrogenlike ions requires to divide the total contribution into two parts: the high energy part (HEP) and the low energy part (LEP). For the evaluation of the HEP, where the virtual photon energy is much larger than the binding energy of the atomic electron, the standard methods of free electron quantum electrodynamics (QED) can be employed (see, for example, [5]). The HEP contribution for the atomic state A looks like (in relativistic units)

$$\Delta E_A^{HEP} = -\frac{e^3}{\pi} \left\{ \frac{1}{3m^3} \left(\ln \frac{m}{\lambda} - \frac{1}{5} \right) (\Delta V)_{AA} + \frac{1}{2m^2} \left(\frac{1}{r} \frac{dV}{dr} (sl) \right) \right\}_{AA} , \tag{55}$$

where m and e is the mass and charge of the electron, V is the Coulomb potential, s is the electron spin, l is the electron orbital momentum and $(...)_{AA}$ denotes the matrix element with the Schrödinger wave functions. The dependence on the "photon mass" λ in Eq. (55) reflects the presence of the infrared divergency in HEP which should be compensated by LEP.

The LEP can be presented in the form [5]

$$\Delta E_A^{LEP} = -\frac{2}{3\pi} \frac{e^2}{m^2} \int_0^{\omega_{\max}} d\omega \omega X_A(E_A - \omega) \, , \tag{56}$$

where

$$X_A(E) = \sum_s \frac{|(\hat{p})_{As}|^2}{E_s - E} \, , \tag{57}$$

$\hat{p} = -i\nabla$ is the electron momentum operator and $\omega_{\max}$ is the cut-off frequency. For the evaluation of X_A the Coulomb Green function can be employed [6]. The use of the spectral expansion (5) yields

$$X_A(E) = \int d\mathbf{r} d\mathbf{r}' \left(\nabla \psi_A^*(\mathbf{r}') G_E(\mathbf{r}'; \mathbf{r}) \nabla \psi_A(\mathbf{r}) \right) \, . \tag{58}$$

Inserting the partial wave expansion (7) in (58) and performing the angular integration we obtain for the ground state $A = 1s$

$$X_A(E) = 4(m\alpha Z)^5 \int_0^\infty dr \int_0^\infty dr' e^{m\alpha Z(r+r')} G_{E1}(r', r) \, . \tag{59}$$

With the use of the representation (24) the expression for $X_{1s}(E)$ takes the form:

$$X_{1s}(E) = 8m(m\alpha Z)^5 \int\limits_1^\infty \frac{d\xi}{\sqrt{\xi^2 - 1}} \left(\frac{\xi + 1}{\xi_1}\right)^\nu$$

$$\times \int\limits_0^\infty dr \int\limits_0^\infty dr' (rr')^{3/2} e^{[m(\alpha Z) + k\xi](r+r')} I_3\left(2k\sqrt{rr'(\xi^2 - 1)}\right) , \tag{60}$$

where $\nu = m\alpha Z/k$, $k = \sqrt{-2mE}$.

The integrals over rr' can be evaluated after expansion of the Bessel's function:

$$I_n^p(a, \gamma) = \int\limits_0^\infty dr \int\limits_0^\infty dr' e^{-a(r+r')} (rr')^p I_n\left(2\gamma\sqrt{rr'}\right)$$

$$= \sum_{s=0}^\infty \frac{\gamma^{2s+n}}{s!(n+s)!} \frac{\Gamma^2\left(p+s+1+\frac{n}{2}\right)}{a^{2\left(p+s+1+\frac{n}{2}\right)}}$$

$$= \frac{\gamma^n \Gamma^2\left(p+1+\frac{n}{2}\right)}{a^{2\left(p+1+\frac{n}{2}\right)} \Gamma(n+1)} {}_2F_1\left(p+1+\frac{n}{2}, p+1+\frac{n}{2}, n+1; \frac{\gamma^2}{a^2}\right) . \tag{61}$$

To evaluate the integral in (60) we should set $n = 3, p = \frac{3}{2}$ in (61). This yields

$$I_3^{3/2}(a, \gamma) = \frac{\gamma^3}{a^8} \Gamma(4) {}_2F_1\left(4, 4, 4; \frac{\gamma^2}{a^2}\right) = \frac{6\gamma^3}{(a^2 - \gamma^2)^4} . \tag{62}$$

Setting then $a = m\alpha Z + k\xi, \gamma = k\sqrt{\xi^2 + 1}$, we obtain

$$X_{1s}(E) = 16(m\alpha Z)^5 k^3 \int\limits_1^\infty d\xi (1 + \xi)^{1+\nu} (1 - \xi)^{1-\nu}$$

$$\times \left[(m\alpha Z)^2 + k^2 + 2m\alpha Z k\xi\right]^{-1} . \tag{63}$$

Comparison of the integral Eq. (63) with the Euler's representation of the hypergeometric function leads to the final result [6]

$$X_{1s}(E) = 128x^5 (1 + x)^{-8} (2 - x)^{-1} {}_2F_1(4, 2 - x, 3 - x; y) , \tag{64}$$

where $x = m\alpha Z/k, y = (1 - x^2)/(1 + x^2)$,

$$\Delta E_{1s}^{LEP} = -128\alpha(\alpha Z)^4 \int\limits_{x_0}^1 dx \frac{1 - x}{(1 + x)^7 (2 - x)} {}_2F_1(4, 2 - x, 3 - x; y) . \tag{65}$$

In (65) we have replaced the integration over ω with the integration over x and set

$$x_0 = \sqrt{\frac{m(\alpha Z)^2}{m(\alpha Z)^2 + 2\omega_{max}}} . \tag{66}$$

The integral in (65) is divergent when $x \to 0$: this is the standard ultraviolet divergency. The divergent terms are proportional to $\frac{1}{x_0^2} \sim \omega_{\max}$ and $\ln x_0 \sim \ln \omega_{\max}$. The former ones vanishes after electron mass renormalization and the latter ones vanishes after matching of the LEP with HEP. This matching can be done with the help of the relation [5]

$$\ln 2\omega_{\min} = \ln \lambda + \frac{5}{6} \tag{67}$$

within the ω region $m\alpha^2 \ll \omega \ll m$. In this region one can put $\omega_{\min} = \omega_{\max}$.

These manipulations finally result in the expression

$$\Delta E_{1s} = \frac{4m\alpha^5 Z^4}{3\pi n^3} \left(\ln \frac{1}{(\alpha Z)^2} - \ln \frac{2K_{10}}{2(\alpha Z)^2} + \frac{19}{30} \right), \tag{68}$$

where K_{10} is Bethe's logarithm for the $1s$ state ($nl = 10$). In the general case

$$\ln \frac{2K_A}{m(\alpha Z)^2} \equiv \frac{\sum_s |(\widehat{p})_{As}|^2 (E_s - E_A) \ln \frac{2|E_s - E_A|}{m(\alpha Z)^2}}{\sum_s |(\widehat{p})_{As}|^2 (E_s - E_A)}. \tag{69}$$

An explicit expression for Bethe's logarithm follows from (65) for the $1s$ state,

$$\ln \frac{2K_{10}}{m(\alpha Z)^2} = 2\ln 2 + \frac{11}{6} - 16 \int_0^1 dx \frac{x(1 - x^2)}{(1 + x)^6 (2 - x)} \, _2F_1(1, 2 - x, 3 - x; y). \tag{70}$$

From (70) the numerical value for Bethe's logarithm can be obtained with any desired accuracy. In case of the $1s$ state $\frac{2K_{10}}{m(\alpha Z)^2} = 19.77$. A review of the latest Lamb shift calculations in the light atoms can be found in [7].

5   Nonresonant Corrections in Atomic Hydrogen

One of the important consequences of the QED theory of the natural line profiles in atoms is the occurrence of the nonresonant (NR) corrections which distort the Lorentz line shape and make the line profile asymmetric [8]. These corrections indicate the limit up to which the concept of the energy of an excited atomic state has a physical meaning - that is the resonance approximation.

The exact theoretical value for the energy of an excited state defined, e.g. by the Green function pole, can be compared directly with measurable quantities only within the resonance approximation, when the line profile is described by the two parameters, energy E and width Γ. Beyond this approximation the evaluation of E and Γ should be replaced by the evaluation of the line profile for the particular process. If the distortion of the Lorentz profile is still small one can formally consider the NR correction as some additional energy shift. Unlike all other energy corrections this correction depends on the particular process which has been employed for the measurement. One can state that the NR corrections set the limit for the accuracy of all atomic frequency standards. Recently NR corrections were evaluated for the Lyman-α $1s - 2p$ transition in the hydrogen atom [9–11]. The process of the resonant Compton scattering was considered as a standard procedure for the determination of the energy levels. For this process the parametric estimate of the NR correction [8] is

$$\delta_{NR} = Cm\alpha^2(\alpha Z)^6 \tag{71}$$

(in relativistic units) where C is a numerical factor. This factor appears to be small ($\sim 10^{-3}$) for the Lyman-α transition [9–11].

Resonance scattering implies that the frequency of the initial photon ω is close to the energy difference $\omega_0 = E_{A'} - E_A$, where $E_{A'}$ is some excited atomic state, A is the initial (ground) state. Within the resonance approximation we retain only one term $n = A'$ in the sum over intermediate states in the amplitude. The Lorentz line profile arises when the we sum up all the electron self-energy insertions in the excited electron state in the resonance approximation [8]. The total cross-section, integrated over the directions of the incident and emitted photons and summed over the polarizations, can be presented in the form

$$\sigma(\omega) = \sigma^{(0)}(\omega) + \sigma^{(1)}(\omega) + \sigma^{(2)}(\omega) , \tag{72}$$

where $\sigma^{(0)}(\omega)$ corresponds to the resonance approximation and is given by the Lorentz formula, $\sigma^{(1)}(\omega)$ represents the interference between the resonant and nonresonant contributions and $\sigma^{(2)}(\omega)$ contains quadratic NR contributions. We assume that the standard way of measuring the resonance frequency is the determination of the maximum in the probability distribution for the given process (the more general procedure is discussed in [10]). In the pure resonance case the maximum condition

$$\frac{d}{d\omega}\sigma^{(0)}(\omega) = 0 \tag{73}$$

corresponds to the resonance frequency value $\omega_{max} = \omega_0$. If we take into account the correction $\sigma^{(1)}(\omega)$ the result will be different, i. e. $\frac{d}{d\omega}\left[\sigma^{(0)}(\omega) + \sigma^{(1)}(\omega)\right] = 0$ with $\omega_{max} = \omega_0 + \delta_{NR}$. The expression for the NR correction looks like [9,10]

$$\delta_{NR} = -\frac{1}{4}\mathrm{Re}\left[\sum_{n \neq A'} \frac{W_{AA;A'n}W_{BB;nA'}}{E_n - E_{A'}} + \sum_{n} \frac{W_{An;A'B}W_{nB;AA'}}{E_n + E_{A'} - 2E_A}\right] , \tag{74}$$

where $W_{AA;A'n}$ is the "mixed" transition probability constructed with the two different transition amplitudes $A \to A'$ and $A \to n$.

The correction δ_{NR} corresponds to the Lyman-α transition if we set $A = 1s$, $A' = 2p$. Employing the partial-wave expansion for the Coulomb Green function one can write the first part of the correction after the angular integration as

$$\delta_1 = -\frac{1}{2}\frac{\alpha^6}{3^7}\int\limits_0^\infty\int\limits_0^\infty dr_1 dr_2 r_1^3 r_2^3 \psi_{1s}^*(r_1)\tilde{G}_{E_{2p}}(r_1;r_2)\psi_{1s}(r_2) , \tag{75}$$

where $\tilde{G}_{E_{2p}}(r_1; r_2)$ is the modified radial Coulomb Green function. Insertion of (40) in (75) and integration over r_1, r_2 results in

$$\delta_1 = -\frac{\alpha^6}{3} \left(\frac{2}{3}\right)^{16} \left(\sum_{m=3}^{\infty} \frac{(m+1)!}{(m-2)(m-2)!} \left({}_2F_1(2-m,5;4;2/3)\right)^2 + 7/2\right). \quad (76)$$

The expansion in (76) converges very fast and for $m = 10$ it gives an error less then 10^{-6}. We obtain the value

$$\delta_{NR}^{(1)} = -2.127209 \cdot 10^{-3} \alpha^6 \text{ a.u.} = -2.1168998 \text{ Hz}.$$

The second term of the correction δ_{NR} can be cast into a form similar to Eq. (75) but with the *ordinary* Coulomb Green function. In this case it is convenient to use the presentation (38) for the corresponding radial Green function. This yields

$$\delta_{NR}^{(2)} = -4\alpha^6 \frac{\nu^7}{(\nu+1)^{10}} \left(\frac{2}{3}\right)^7$$

$$\times \sum_{m=2}^{\infty} \frac{(m+1)m(m-1)}{m-\nu} \left({}_2F_1(2-m,5;4;2/(\nu+1))\right)^2. \quad (77)$$

Retaining only six terms of the expansion (77) yields an accuracy of 10^{-6}. Thus for the second term we have

$$\delta_{NR}^{(2)} = -0.821625 \cdot 10^{-3} \text{ a.u.} = -0.81764337 \text{ Hz}.$$

In [10,11] it was found that the quadratic NR contribution to the total cross-section from $2p_{3/2}$ state is also important. The enhancement follows from the small energy denominator $\Delta E_f = E_{2p_{3/2}} - E_{2p_{1/2}}$. This contribution to the interference term $\sigma^{(1)}(\omega)$ vanishes after the angular integration. However it survives in $\sigma^{(2)}(\omega)$ and is given by the formula

$$\delta_{NR}^{(3)} = \frac{1}{8} \frac{\Gamma_{2p}^4}{\Delta E_f^3}, \quad (78)$$

where Γ_{2p} is the width of the level $2p$ (we neglect the difference between $\Gamma_{2p_{1/2}}$ and $\Gamma_{2p_{3/2}}$).

A parametrical estimate gives the result

$$\delta_{NR}^{(3)} = C_3 m\alpha^4 (\alpha Z)^4, \quad (79)$$

which coincides with (71) for the $Z = 1$. The numerical value is

$$\delta_{NR}^{(3)} = 0.00948\,\alpha^6 \text{ a.u.} = 9.42 \text{ Hz}.$$

Finally, an "asymmetry" correction of the same order arises from the resonant term when we replace the width $\Gamma(\omega_0)$ by $\Gamma(\omega)$. In case of the Lyman-α

transition $\Gamma(\omega_0) = \left(\frac{2}{3}\right)^8 \alpha^3$ a.u. should be replaced by $\Gamma(\omega) = \frac{2^{11}}{3^9}\alpha^3\omega$ a.u. The additional "asymmetry" shift is $\delta_{AS} = 1.015 \cdot 10^{-3}\alpha^3 = 1.007$ Hz.

The total correction

$$\delta = \delta_{NR}^{(1)} + \delta_{NR}^{(2)} + \delta_{NR}^{(3)} + \delta_{AS} = 7.50 \text{ Hz} \tag{80}$$

can be considered as an absolute limit for the accuracy of the Lyman-α transition frequency measurements.

As it was already told in the beginning of this section this correction only formally looks like an energy correction. Actually NR correction presents the difference between the theoretical value for the transition frequency and the frequency that corresponds to the maximum of the distorted Lorentz curve. The deduction of the transition frequency from the asymmetrical Lorentz line profile becomes ambiguous and the improvement of its accuracy beyond the limit given by (30) is impossible.

6   Relativistic Coulomb Green Function

In the relativistic theory of the hydrogen atom and especially in the highly charged hydrogen-like ions where relativistic effects are important, the relativistic Coulomb Green function (RCGF) is widely exploited. This function is defined by the equation

$$\left(\hat{H}_D - E\right)\tilde{G}_E\left(\boldsymbol{r};\boldsymbol{r}'\right) = \delta\left(\boldsymbol{r} - \boldsymbol{r}'\right) , \tag{81}$$

where $\hat{H}_D$ is the Dirac Hamiltonian

$$\hat{H}_D = \boldsymbol{\alpha}\hat{\boldsymbol{p}} + \beta m - eV . \tag{82}$$

Here, α, β are the Dirac matrices, $\hat{\boldsymbol{p}} \equiv -i\nabla$ is the electron momentum operator, $V \equiv -\frac{eZ}{r}$ is the Coulomb interaction with the nucleus, m and e are the electron mass and charge. In (82) relativistic units are used.

The spectral expansion similar to (5) can be employed for the RCGF. This expansion does not differ formally from (5),

$$\tilde{G}_E\left(\boldsymbol{r};\boldsymbol{r}'\right) = \sum_i \frac{\psi_i^\dagger(\boldsymbol{r})\psi_i(\boldsymbol{r}')}{E_i - E} . \tag{83}$$

However, in (83) the sum is extended over the total spectrum of the Dirac Hamiltonian $\hat{H}_D$. The function ψ_i is the one-column 4-matrix with the components $\psi_{i\alpha}$ where α are the spinor indices,

$$\psi_i(\boldsymbol{r}) = \begin{pmatrix} \varphi_1(\boldsymbol{r}) \\ \varphi_2(\boldsymbol{r}) \\ \chi_1(\boldsymbol{r}) \\ \chi_2(\boldsymbol{r}) \end{pmatrix} . \tag{84}$$

This is the bispinor form of the Dirac wave function. Another form is also often used,

$$\psi_i(\boldsymbol{r}) = \begin{pmatrix} \varphi_i(\boldsymbol{r}) \\ \chi_i(\boldsymbol{r}) \end{pmatrix} , \tag{85}$$

where φ, χ are the "upper" and "lower" spinors. The Hermitian conjugated wave function $\psi_i^\dagger(\boldsymbol{r})$ is presented by the one-row 4-matrix

$$\psi_i^\dagger(\boldsymbol{r}) = (\varphi_1^*(\boldsymbol{r})\varphi_2^*(\boldsymbol{r})\chi_1^*(\boldsymbol{r})\chi_2^*(\boldsymbol{r})) , \tag{86}$$

Hence, the Green function is presented by 4×4 matrix with spinor indices.

The different methods were applied for the explicit evaluation of RCGF. In particular, the expansion in the parameter αZ (α is the fine structure constant) was obtained in [12]. The first term of this expansion coincides with the nonrelativistic CGF.

The partial wave expansion similar to (7) was elaborated in [13], [14].

For the construction of the partial wave expansion it is more convenient to use the function $G_E = \beta \tilde{G}_E$. This function satisfies the equation

$$\left[\beta\boldsymbol{\alpha}\hat{\boldsymbol{p}} - \beta\left(E + \frac{\alpha Z}{r}\right) + m\right] G_E(\boldsymbol{r};\boldsymbol{r}') = \delta(\boldsymbol{r} - \boldsymbol{r}') \tag{87}$$

and can be presented in the form

$$G_E(\boldsymbol{r};\boldsymbol{r}') \equiv \left[-\beta\boldsymbol{\alpha}\hat{\boldsymbol{p}} + \beta\left(E + \frac{\alpha Z}{r} + m\right)\right] \Phi_E(\boldsymbol{r};\boldsymbol{r}') . \tag{88}$$

The function $\Phi_E(\boldsymbol{r};\boldsymbol{r}')$ satisfies the quadric equation that arises after the insertion of (88) in (87):

$$\left[-\Delta + m^2 - E^2 - \frac{2E\alpha Z}{r} - \frac{(\alpha Z)^2}{r^2} - i\boldsymbol{\alpha}\boldsymbol{n}\frac{\alpha Z}{r^2}\right] \Phi_E(\boldsymbol{r};\boldsymbol{r}') = \delta(\boldsymbol{r} - \boldsymbol{r}') . \tag{89}$$

Here $\boldsymbol{n} \equiv \boldsymbol{r}/r$. Separating the radial and angular parts of the Laplacian we can rewrite (89) in the form:

$$\left[-\frac{1}{2}\frac{\partial}{\partial r}r^2\frac{\partial}{\partial r} + m^2 - E^2 - \frac{2E\alpha Z}{r} + \frac{\hat{N}}{r^2}\right] \Phi_E(\boldsymbol{r};\boldsymbol{r}') = \delta(\boldsymbol{r} - \boldsymbol{r}') , \tag{90}$$

where the angular operator $\hat{N}$ is

$$\hat{N} \equiv \hat{\boldsymbol{l}}^2 - (\alpha Z)^2 - i\alpha Z\boldsymbol{\alpha}\boldsymbol{n} \tag{91}$$

and $\hat{\boldsymbol{l}}$ is the orbital angular momentum operator. The operator $\hat{N}$ can be presented also in the form

$$\hat{N} = \begin{pmatrix} \hat{K}^2 + \hat{K} - (\alpha Z)^2 & -i\alpha Z\boldsymbol{\sigma}\boldsymbol{n} \\ -i\alpha Z\boldsymbol{\sigma}\boldsymbol{n} & \hat{K}^2 + \hat{K} - (\alpha Z)^2 \end{pmatrix} , \tag{92}$$

where $\boldsymbol{\sigma}$ are the Pauli matrices and the two-component operator $\hat{K} = -(1 + \boldsymbol{\sigma}\boldsymbol{l})$ possess the properties

$$\hat{K}\Omega_{jlM} = \kappa_{jl}\Omega_{jlM} \,, \tag{93}$$

$$\hat{K}\Omega_{j\bar{l}M} = -\kappa_{jl}\Omega_{j\bar{l}M} \,.$$

In (93) the functions Ω_{jlM} are spherical spinors

$$\Omega_{jlM}(\boldsymbol{n}) = \sum_{m\mu} C^{l\frac{1}{2}}_{jM}(m\mu)Y_{lm}(\boldsymbol{n})\eta_\mu \,, \tag{94}$$

Y_{lm} are spherical functions, η_μ are two-component spinors, and $C^{l\frac{1}{2}}_{jM}(m\mu)$ are Clebsh-Gordan coefficients. The notation $\bar{l} = -l$ is used in (93) and the angular quantum numbers κ_{jl} are defined as $\kappa_{jl} = \pm\left(j + \frac{1}{2}\right)$ for $j = l \mp \frac{1}{2}$.

The eigenfunctions and eigenvalues for the operator $\hat{N}$ look like

$$\hat{N}P_\gamma(\boldsymbol{n};\boldsymbol{n}') = \gamma(\gamma+1)P_\gamma(\boldsymbol{n};\boldsymbol{n}') \,, \tag{95}$$

where

$$\gamma^2 = \kappa^2 - (\alpha Z)^2 \tag{96}$$

and

$$P_\gamma(\boldsymbol{n};\boldsymbol{n}') = \sum_{jlM} \begin{pmatrix} \frac{\gamma+\kappa}{2\gamma}\Omega_{jlM}(\boldsymbol{n})\Omega^\dagger_{jlM}(\boldsymbol{n}') & \frac{i\alpha Z}{2\gamma}\Omega_{jlM}(\boldsymbol{n})\Omega^\dagger_{j\bar{l}M}(\boldsymbol{n}') \\ \frac{i\alpha Z}{2\gamma}\Omega_{j\bar{l}M}(\boldsymbol{n})\Omega^\dagger_{jlM}(\boldsymbol{n}') & \frac{\gamma+\kappa}{2\gamma}\Omega_{j\bar{l}M}(\boldsymbol{n})\Omega^\dagger_{j\bar{l}M}(\boldsymbol{n}') \end{pmatrix} \,. \tag{97}$$

The possible values of κ were defined above. In (95) the vector $\boldsymbol{n}$ plays the role of the argument of the function P_γ and the vector $\boldsymbol{n}'$ should be considered as parameter.

The functions $P_\gamma(\boldsymbol{n};\boldsymbol{n}')$ satisfy the completeness relation

$$\sum_\gamma P_\gamma(\boldsymbol{n};\boldsymbol{n}') = \delta(\boldsymbol{n} - \boldsymbol{n}') \tag{98}$$

which follows from the completeness relation for the spherical spinors.

Using (98) we can rewrite (8) in the form

$$\delta(\boldsymbol{r} - \boldsymbol{r}') = \frac{1}{rr'}\delta(r - r')\sum_\gamma P_\gamma(\boldsymbol{n};\boldsymbol{n}') \,. \tag{99}$$

We can also expand the function $\Phi_E(\boldsymbol{r};\boldsymbol{r}')$ in the complete set of the angular functions $P_\gamma(\boldsymbol{n};\boldsymbol{n}')$:

$$\Phi_E(\boldsymbol{r};\boldsymbol{r}') = \frac{1}{rr'}\sum_\gamma \Phi_{\gamma E}(r;r') P_\gamma(\boldsymbol{n};\boldsymbol{n}') \,. \tag{100}$$

Then the insertion of (99) and (100) in (90) leads to the following equations for the radial functions $\Phi_{\gamma E}(r;r')$:

$$\left[-\frac{d^2}{dr^2} + m^2 - E^2 - \frac{2E\alpha Z}{r} + \frac{\gamma(\gamma+1)}{r^2} \right] \Phi_{\gamma E}(r;r') = \delta(r - r') . \tag{101}$$

Equation (101) coincides with the (9) if we set in (9) $l \to \gamma, E \to E^2 - m^2, Z \to 2E\alpha Z$. Then we obtain for $\Phi_{\gamma E}$ the solution similar to (17)

$$\Phi_{\gamma E}(r;r') = \frac{\nu}{2E\alpha Z} \frac{\Gamma(\gamma+1-\nu)}{\Gamma(2\gamma+2)}$$

$$\times M_{\nu,\gamma+\frac{1}{2}} \left(\frac{4E\alpha Z}{\nu} r_< \right) W_{\nu,\gamma+\frac{1}{2}} \left(\frac{4E\alpha Z}{\nu} r_> \right) \tag{102}$$

where $\nu = \frac{2E\alpha Z}{\sqrt{2(m^2-E^2)}}$. Due to the factor $\Gamma(\gamma+1-\nu)$ the function $\Phi_{\gamma E}(r;r')$ as a function of the complex variable E possess poles, corresponding to the Sommerfeld energy values

$$E_{n_r jl} = m \left[1 + \frac{(\alpha Z)^2}{\left(\sqrt{\kappa_{jl}^2 - (\alpha Z)^2} + n_r^2 \right)^2} \right]^{-\frac{1}{2}} . \tag{103}$$

Here the radial quantum number n_r runs over the values

$$n_r = \begin{bmatrix} 0, 1, 2, ... & \text{if } \kappa < 0 , \\ 1, 2, 3, ... & \text{if } \kappa > 0 . \end{bmatrix} \tag{104}$$

Still the expression (102) does not solve the problem completely since we have to obtain from the solution of the quadric equation (89) the solution of the initial equation (87). For this purpose we insert the expansion (100) in (88). This yields

$$G_E(r;r') = \sum_\gamma \left[i\frac{d}{dr} \left(\frac{1}{rr'} \Phi_{\gamma E}(r;r') \right) \beta(\alpha n) P_\gamma(n;n') \right. \tag{105}$$

$$\left. - \frac{1}{rr'} \Phi_{\gamma E}(r;r') \left[\beta(\alpha \hat{p}) + (E + \frac{\alpha Z}{r})\beta + m \right] P_\gamma(n;n') \right] .$$

It remains to define the action of the operators (αn) and (αp) on the functions $P_\gamma(n;n')$. It can be done using of the action of these operators on the spherical spinors:

$$\beta(\alpha n) P_\gamma(n;n') = \tag{106}$$

$$\sum_{jlM} \left(\begin{array}{cc} -\frac{i\alpha Z}{2\gamma} \Omega_{jlM}(n)\Omega_{jlM}(n') & -\frac{\gamma-\kappa}{2\gamma}\Omega_{jlM}(n)\Omega_{jlM}^\dagger(n') \\ \frac{\gamma+\kappa}{2\gamma} \Omega_{j\bar{l}M}(n)\Omega_{j\bar{l}M}^\dagger(n') & \frac{i\alpha Z}{2\gamma}\Omega_{j\bar{l}M}(n)\Omega_{j\bar{l}M}^\dagger(n') \end{array} \right) ,$$

$$\beta(\alpha n) P_\gamma(n;n') = \tag{107}$$

$$\frac{1}{r} \sum_{jlM} \left(\begin{array}{cc} -\frac{i\alpha Z(1-\kappa)}{2\gamma} \Omega_{jlM}(n)\Omega_{jlM}^\dagger(n') & -\frac{\gamma-\kappa}{2\gamma}(1-\kappa)\Omega_{jlM}(n)\Omega_{jlM}^\dagger(n') \\ -\frac{\gamma+\kappa}{2\gamma}(1+\kappa)\Omega_{j\bar{l}M}(n)\Omega_{j\bar{l}M}^\dagger(n') & -\frac{i\alpha Z}{2\gamma}(1+\kappa)\Omega_{j\bar{l}M}(n)\Omega_{j\bar{l}M}^\dagger(n') \end{array} \right) .$$

The details of the RCGF theory, including series representation for the radial part and the construction of the modified RCGF, can be found in [15]. The applications of the RCGF are numerous, but the most important results amongst them (electron self-energy and vacuum polarization calculations) only have been obtained numerically [16].

In the next section we will describe one of the simplest applications of RCGF which can be presented explicitly in the analytic form – the relativistic polarizability of H-like ions in a static electric field [17,18].

7   Relativistic Polarizability of the H-Like Ions

We consider an one-electron ion in its ground state and in an static external electric field. The second order Stark shift for the state A is given by the formula

$$\Delta E_A^{(2)} = \alpha F^2 \sum_n \frac{|<A|z|n>|^2}{E_n - E_A} , \qquad (108)$$

where F is the field strength and the field is oriented along z axis. The summation in (108) is extended over the total Dirac spectrum, $\langle n|$ and E_n are the Dirac eigenfunctions and eigenvalues for the electron in the field of nucleus.

The polarizability of the ion in the state A is defined by the relation

$$\Delta E_A^{(2)} = -2\beta_A F^2 . \qquad (109)$$

Due to the spatial parity conservation $\langle n|z|n\rangle = 0$ and the term $n = A$ is absent in the sum over n in Eq. (108). For the evaluation of this sum the RCGF in the form (83) can be used. Taking the definition of G_E into account we can write

$$\Delta E_A^{(2)} = \alpha F^2 \lim_{E \to E_A} \int \int d\boldsymbol{r_1} d\boldsymbol{r_2} \psi_A^\dagger(\boldsymbol{r_1}) \beta z_2 G_E(\boldsymbol{r_1}; \boldsymbol{r_2}) \beta z_1 \psi_A^\dagger(\boldsymbol{r_1}) . \qquad (110)$$

Inserting the expression (105) in (110), using (106) and (107) and performing the angular reduction we obtain

$$\Delta E_A^{(2)} = \frac{\alpha F^2}{9} \lim_{E \to E_A} \left[M_{1/2}^1 \left(E, 1\frac{1}{2}0 \right) + 2M_{3/2}^1 \left(E, 1\frac{1}{2}0 \right) \right] . \qquad (111)$$

Here

$$M_J^L (E, njl) = R_{JL}^{(1)} (E, njl, njl) \qquad (112)$$
$$+ R_{JL}^{(2)} \left(E, njl, nj\bar{l} \right) + R_{JL}^{(3)} \left(E, nj\bar{l}, njl \right) + R_{JL}^{(4)} \left(E, nj\bar{l}, nj\bar{l} \right) ,$$

$$R_{JL}^{(i)} (E, njl, n'j'l') = \int r_1^3 dr_1 r_2^3 dr_2 g_{njl}(r_1) G_{JL}^{(j)} (E, r_1, r_2) g_{n'j'l'}(r_2) , \qquad (113)$$

g_{njl} is the upper radial component of the Dirac wave function for the electron in the Coulomb field,

$$G_{JL}^{(1)} (E, r, r') = \frac{2m}{\nu} (1 + \epsilon) \left[-(\alpha \nu Z + \kappa) S_{JL}^{(1)} \right. \qquad (114)$$

$$+ (\alpha\nu Z - \kappa - 2\gamma - 2\eta)S_{JL}^{(0)} - S_{JL}^{+}\Big] \, ,$$

$$G_{JL}^{(2)}(E, r, r') = \frac{2m}{\nu}\sqrt{1 - \epsilon^2}\left[(\alpha\nu Z + \kappa)S_{JL}^{(1)} - (\alpha\nu Z - \kappa)S_{JL}^{(0)} - S_{JL}^{(-)}\right] \, ,$$

$$G_{JL}^{(3)}(E, r, r') = -\frac{2m}{\nu}\sqrt{1 - \epsilon^2}\left[(\alpha\nu Z + \kappa)S_{JL}^{(1)} - (\alpha\nu Z - \kappa)S_{JL}^{(0)} - S_{JL}^{(-)}\right] \, ,$$

$$G_{JL}^{(4)}(E, r, r') = -\frac{2m}{\nu}(1 - \epsilon)$$

$$\times \left[-(\alpha\nu Z + \kappa)S_{JL}^{(1)} - (\alpha\nu Z - \kappa + 2\gamma + 2\eta)S_{JL}^{(0)} - S_{JL}^{(+)}\right] \, ,$$

$$\epsilon = \frac{E}{m}, \, \nu = (1 - \epsilon)^{-\frac{1}{2}}, \, \eta = \alpha\nu Z\epsilon \, ,$$

$$S_{JL}^{(i)}(r, r') = (\rho\rho')^{\gamma-1}\exp\left[-(\rho + \rho')/2\right] \tag{115}$$

$$\times \sum_{k=0}^{\infty} \frac{k!L_k^{2\gamma}(\rho)L_k^{2\gamma}(\rho')}{\Gamma(k + 2\gamma + 1)(k + \gamma - \eta + i)} \, ,$$

$$S_{JL}^{(\pm)}(E, r, r') = (\rho\rho')^{\gamma-1}\exp\left[-(\rho + \rho')/2\right] \times \tag{116}$$

$$\times \sum_{k=0}^{\infty} \frac{k!\left[L_k^{2\gamma-1}(\rho)L_k^{2\gamma}(\rho') \pm L_k^{2\gamma}(\rho)L_k^{2\gamma-1}(\rho')\right]}{\Gamma(k + 2\gamma)(k + \gamma - \eta)} \, ,$$

$i = 1, 2, \quad \rho = 2r/\nu, \rho' = 2r'/\nu, \, L_n^m(x)$ are the Laguerre polynomials.

Evaluation of the integrals in (111) for the ground state $A \equiv 1s_{1/2}$ with help of the formula

$$\int\limits_0^\infty dx\exp(-x)x^\alpha L_n^\beta(x) = [\Gamma(\alpha + 1)\Gamma(\beta - \alpha + n)\Gamma(\beta - \alpha)n!]^{-1} \tag{117}$$

in the limit $E \to E_A$ yields

$$\Delta E_A^{(2)} = -\frac{\alpha m}{72(\alpha Z)^4}\left[2(2\gamma + 1)^3 + \gamma_1(3 - \gamma_1^2)(2\gamma_1 + 1)^2 + 4(2\gamma_1 + 1)\right. \tag{118}$$

$$+ \frac{4\Gamma(\gamma_1 + \gamma_2 + 2)^2}{\Gamma(2\gamma_1 + 1)}\left(\frac{1}{\Gamma(2\gamma_2 + 1)}\left(2\gamma_1 + \frac{3}{\gamma_2 - \gamma_1} - \frac{\gamma_1^2}{\gamma_2 - \gamma_1 + 1}\right)\right)$$

$$+ \frac{1}{[\Gamma(\gamma_2 - \gamma_1 - 1)]^2}\sum_{n=1}^\infty \frac{[\Gamma(\gamma_2 - \gamma_1 + n - 1)]^2}{\Gamma(n + 2\gamma_2 + 1)n!}$$

$$\times \left(\frac{3 - 2\gamma_1(\gamma_1 + \gamma_2)}{n + \gamma_2 - \gamma_1} - \frac{\gamma_1^2}{n + \gamma_2 - \gamma_1 + 1}\right)$$

$$+ \frac{2\gamma_1}{\Gamma(\gamma_2 - \gamma_1 - 1)\Gamma(\gamma_2 - \gamma_1 - 2)}$$

$$\left.\times \sum_{n=1}^\infty \frac{\Gamma(\gamma_2 - \gamma_1 + n - 1)\Gamma(\gamma_2 - \gamma_1 + n - 2)}{\Gamma(n + 2\gamma_2)n!\Gamma(n + \gamma_2 - \gamma_1)}\right)\right] \, ,$$

where γ_κ is defined by (96).

The expression (118) can be expanded in powers of the parameter αZ. The first two terms of this expansion for the polarizability are

$$\beta_{1s_{1/2}} = \alpha m (\alpha Z)^{-4} \left[\frac{9}{2} + \frac{14}{3}(\alpha Z)^2 \right] . \tag{119}$$

The first term in (119) is the nonrelativistic value of the polarizability of H-like ion and the second term is the lowest-order relativistic correction.

Acknowledgments

This work was supported by the RFBR grants 02-02-16379, 02-02-06689-mas and by the Minobrazovanje grant E00-3.1-7.

References

1. L. Hostler: J. Math. Phys. **5**, 591(1964)
2. V.A. Fock: *Nachala kvantovoj mechaniki (Foundations of quantum mechanics)* (Moscow, Nauka, 1976) (in Russian)
3. L.P. Rapoport, B.A. Zon and N.L. Manakov: *Teorija mnogofotonnych prozessov v atomach. (Theory of the multiphoton processes in atoms)* (Moscow, Atomizdat, 1978) (in Russian)
4. B.A. Zon, N.L. Manakov and L.P. Rapoport: Zh. Eksp. Teor. Fiz **55**, 924 (1968)
5. A.I. Akhiezer and V.B. Berestetskii: *Quantum Electrodynamics* (New York, Wiley Interscience 1965)
6. Ya.I. Granovskii: Zh. Exsp. Teor. Fiz **56**, 605 (1969) [Engl. Transl. Sov. Phys. JETP **29**, 333 (1969)]
7. M.I. Eides, H. Grotch, and V.A. Shelyuto: Phys. Rep. **342**, 63(2001)
8. F. Low: Phys. Rev. **88**, 53 (1951)
9. L.N. Labzowsky, D.A. Solovyev, G. Plunien and G. Soff: Phys. Rev. Lett. **87**, 143003 (2001)
10. U.D. Jentschura and P.J. Mohr: Can. J. Phys. **80**, 633 (2002)
11. L.N. Labzowsky, D.A. Solovyev, G. Plunien and G. Soff: Phys. Rev. A**65**, 054502 (2002)
12. V.G. Gorshkov: Zh. Eksp. Teor. Fiz. **47**, 352 (1964) [Engl. Transl. Sov. Phys. JETP-20, 234 (1964)]
13. E.H. Wichmann and N.M. Kroll: Phys. Rev. **101**, 843 (1956)
14. P.C. Martin and R.J. Glauber: Phys. Rev. **109**, 1307 (1958)
15. L. Labzowsky, G. Klimchitskaya and Yu. Dmitriev: *Relativistic Effects in the Spectra of Atomic Systems.* IOP Publishing, Bristol and Philadelphia, 1993
16. P.J. Mohr, G. Plunien and G. Soff: Physics Reports **293**, 227 (1998)
17. L. Labzowsky: Vestnik Leningradskogo Universiteta, Ser. fiz. (Communications of the Leningrad University, Physics) **10**, 19 (1972) (in Russian)
18. N.L. Manakov and S.A Zapryagaev: Phys. Lett. **58A**, 23 (1976)

Part II

Muonic and Exotic Atoms and Nuclear Effects

Atomic Cascade and Precision Physics with Light Muonic and Hadronic Atoms

Thomas S. Jensen[1,2] and Valeri E. Markushin[1]

[1] Paul Scherrer Institute, CH-5232 Villigen PSI, Switzerland
[2] Institut für Theoretische Physik der Universität Zürich, Winterthurerstrasse 190, CH-8057 Zürich, Switzerland

Abstract. Recent progress in the theory of atomic cascade in exotic hydrogen atoms is reviewed from the viewpoint of precision experiments with $\mu^- p$, $\pi^- p$, $K^- p$, and $\bar{p}p$ atoms.

PACS Nos.: 36.10.-k, 32.30.Rj, 32.70.Jz, 32.80.Cy

1 Introduction

Exotic atoms provide numerous opportunities to investigate nuclear properties and low–energy nuclear reactions in well defined and controlled conditions. The properties of atomic states can be calculated with high precision from QED, while the nuclear interaction in exotic atoms can be described with a special kind of perturbation theory based on the hierarchy of scales [1–4]: the range of the nuclear interaction is much smaller then the characteristic atomic scale and the nuclear energy scale is much larger then the atomic one. The nuclear interaction usually results in small corrections to the energies of atomic states; nevertheless, due to extremely high precision of spectroscopic methods, some of these nuclear effects can be measured with accuracy of 1% or better. Of particular interest are the cases where this level of accuracy, which is often considered as a precision measurement in nuclear physics, cannot not be reached with other methods.

There are both experimental and theoretical challenges in precision spectroscopy of exotic atoms. The proton charged radius can be determined with an accuracy of 10^{-3} from the energy splitting between the $2S$ and $2P$ states in muonic hydrogen. This goal is expected to be reached in the muonic hydrogen Lamb shift experiment [5,6] presently in progress at PSI. The feasibility of this experiment depends crucially on the population and the lifetime of the metastable $2S$ state of $\mu^- p$, and a good understanding of the atomic cascade is essential for performing this measurement. The πN scattering length can be determined with an accuracy better than 1% by measuring the nuclear shifts and widths of the K X-ray lines in pionic hydrogen. A significant improvement of the earlier measurements [7] is expected in the new experiment at PSI [8,9]. The ultimate accuracy of the nuclear width is determined by the Doppler broadening corrections to the line profile, that must be calculated in a reliable cascade model. Similar problems exist in the spectroscopy of other hadronic atoms. In the case of kaonic hydrogen, theoretical predictions for the relative yields of the individual K lines can be used to improve the determination of the KN scattering lengths which are obtained using X-ray spectroscopy [10,11]. The $2P$ nuclear widths of antiprotonic hydrogen can be determined from the L X-ray spectra [12,13], and the Doppler broadening corrections turn out to be important for precision measurements.

At the same time, these precision spectroscopic experiments allow one to get detailed information about the kinetics of atomic cascade, that was not available in the past. On the theoretical side, the ultimate goal is the *ab initio* cascade calculations. By confronting the theoretical predictions with the experimental data one can verify whether the cascade model is complete and all important cascade processes are adequately treated. In this review we shall focus our attention on the hydrogen–like exotic atoms with $Z = 1$, which, despite of their simple structure, exhibit very complicated "life histories" reflecting a broad variety of reactions occurring during the cascade, and discuss the cascade features that are important for the feasibility of the above mentioned precision experiments.

The paper is organized as follows. Section 2 gives a brief introduction into the present state of the theory of atomic cascade in exotic hydrogen atoms. The cascade problems related to the precision spectroscopy of muonic hydrogen are discussed in Sect. 3. Precision measurements of the pion–nucleon scattering length using the pionic hydrogen spectroscopy are considered in Sect. 4. The atomic cascades in kaonic and antiprotonic hydrogen are discussed in Sects. 5 and 6. The results are summarized in Sect. 7.

2   Atomic Cascade

The traditional and most common way to produce exotic atoms is to stop low energetic heavy negative particles (μ^-, π^-, K^-, $\bar{p}$, etc.) in an appropriate target. After slowing down the negative particles are captured into atomic orbits with the characteristic size of the electron cloud. Ingenious devices, such as magnetic bottles and cyclotron traps (see [14,15] and references therein), have been developed to control the location of the formed exotic atoms and to significantly improve the signal–to–background ratio and the resolution of X-ray detectors. Exotic hydrogen atoms are usually studied in hydrogen or deuterium targets, which may contain some small admixture of other elements, e.g. for calibration purposes. For the sake of simplicity, we restrict our discussion to the case of pure hydrogen target.

Exotic hydrogen atoms are initially formed in highly excited states with the principal quantum number $n \sim \sqrt{m_x/m_e}$ where m_x is the reduced mass of the exotic atom $x^- p$ and m_e is the electron mass [16–18]. The exotic atoms deexcite in a sequence of radiative and collisional processes forming a so–called atomic cascade, see Fig. 1a and Table 1. The radiative deexcitation has the well–known properties of the electrodipole transitions: the change of the orbital angular momentum l satisfies the selection rule $|l_i - l_f| = 1$ and the rates drop fast for increasing initial quantum number n_i, see Fig. 1b. An almost pure radiative cascade is possible only at very low target density (less than 0.05 mbar for

a   **b**

$$n_i \approx (m_\pi / m_e)^{1/2}$$

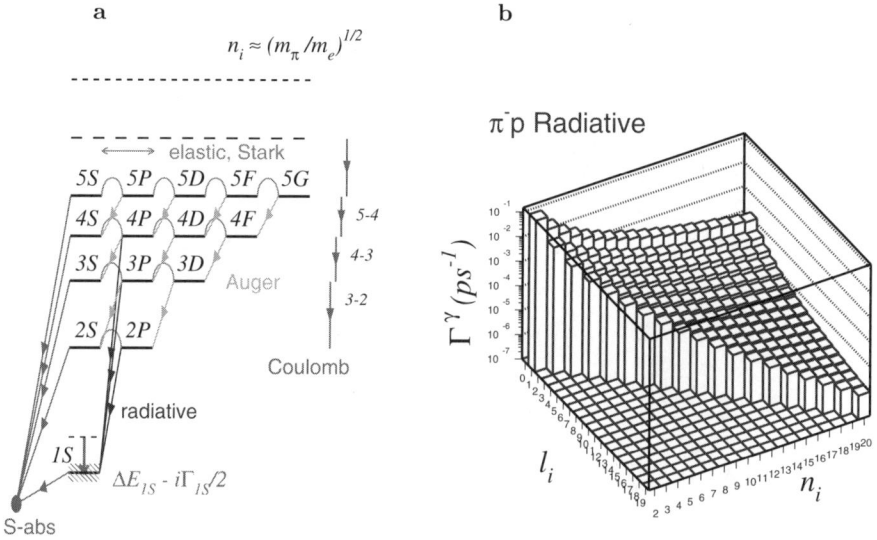

Fig. 1. a A simplified scheme of the atomic cascade in pionic hydrogen. See Table 1 for the description of the cascade processes. **b** The total radiative deexcitation rates in pionic hydrogen *vs.* the initial state $n_i l_i$

Table 1. Cascade processes in exotic hydrogen atoms

Process	Example	Reference
Radiative:	$(x^- p)_{n_i l_i} \to (x^- p)_{n_f l_f} + \gamma$	[19]
Stark transitions:	$(x^- p)_{nl_i} + H_2 \to (x^- p)_{nl_f} + H_2^*$	[16,20–27]
External Auger effect:	$(x^- p)_i + H \to (x^- p)_f + p + e^-$	[16,27,28]
Coulomb transitions:	$(x^- p)_{n_i} + H_2 \to (x^- p)_{n_f} + H + H, \ n_f < n_i$	[25,27,29,30]
Elastic scattering:	$(x^- p)_{nl} + H_2 \to (x^- p)_{nl} + H_2^*$	[26,27,31,32]
Absorption:	$(\pi^- p)_i + H \to \pi^0 + n + H$	[16,24–26]
Nuclear reaction:	$(\pi^- p)_{ns} \to \pi^0 + n, \ \gamma + n$	[16]
Weak decay:	$\pi^- \to \mu^- \bar\nu_\mu$	

$\mu^- p)$ that is seldom realized in practice. At least three collisional mechanisms are essential for the basic understanding of the atomic cascade: the Stark mixing, the external Auger effect, and the Coulomb deexcitation, see Fig. 2. The rates of the collisional processes are proportional to the target density, At liquid hydrogen density, the collisional deexcitation dominates over the radiative transitions through the whole cascade except for the lowest part with the initial states $n_i = 2.3$.

The Stark mixing [16,26] results in the transitions among the l–sublevels with the same n. This is a very fast collisional process because the exotic hydrogen atoms, which are electroneutral and have no electrons, can easily pass through

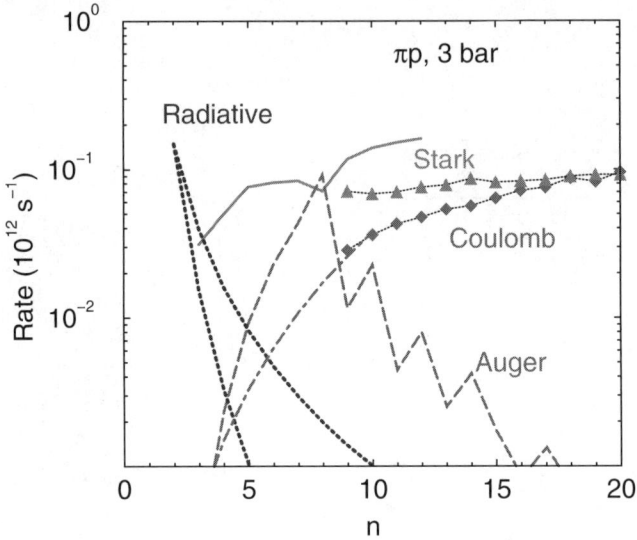

Fig. 2. The *l*-average rates for pionic hydrogen with the lab kinetic energy $T = 1$ eV in gaseous target at 3 bar. The Coulomb deexcitation (filled diamonds) and Stark mixing (filled triangles) rates calculated in the CTMC model [27] are shown in comparison with the results of the semiclassical fixed field model for Stark mixing (*solid line*) and Auger deexcitation (light *dashed line*) [26,27]. The *dash-dotted line* shows the Coulomb deexcitation rate obtained by scaling the result from [29] for muonic hydrogen, see [27]. The radiative rates $nP \rightarrow 1S$ and $n(n-1) \rightarrow (n-1)(n-2)$ are shown with *dotted lines*

the regions of strong electric field inside ordinary atoms. The external Auger effect [16,27] is the deexcitation process in which the transition energy is mainly carried away by the electron. The Auger mechanism favors the transitions with the minimal change of the principal quantum number, that is sufficient for the ionization of the hydrogen molecule. The rates of the Auger transitions reach their maximum when the transitions with $n_f = n_i - 1$ becomes energetically possible, then the Auger deexcitation rates rapidly decrease with decreasing the initial principal quantum number n_i, The Coulomb deexcitation [27,29,30] is the dominant deexcitation mechanism at the initial stage of the atomic cascade. An important feature of the Coulomb mechanism is that the transition energy is shared by the recoiling particles (the $x^- p$ and the two hydrogen atoms) having comparable masses, therefore the Coulomb deexcitation can lead to significant acceleration of exotic atoms during the cascade. The deceleration of the exotic atoms takes place in the elastic and Stark collisions. In most cases the exotic atoms are epithermal during the cascade, with the characteristic kinetic energy being much higher than the target temperature. A typical example of the *n*-dependence of the cascade rates in shown in Fig. 2 for pionic hydrogen in gaseous target at 3 bar.

Table 2. Examples of cascade models. The $+/-$ signs indicate whether the evolution of the l–distribution and the kinetic energy distribution $w_{nl}(T)$ is taken into account

Model	l	$w_{nl}(T)$	References	Atoms
Leon, Bethe (1962)	$-$	$-$	[16]	$\pi^- p,\, K^- p$
Borie, Leon (1980)	$+$	$-$	[20]	$\pi^- p,\, \mu^- p,\, K^- p,\, \bar{p}p$
Markushin (1981)	$+$	$-$	[21]	$\mu^- p,\, \mu^- d$
Reifenröther, Klempt (1989-90)	$+$	$-$	[35,36]	$\pi^- p,\, K^- p,\, \bar{p}p$
Czaplinski et al. (1994)	$+$	$+$	[37]	$\mu^- p,\, \mu^- d$
Markushin (1994)	$+$	$+$	[33]	$\mu^- p,\, \mu^- d$
Aschenauer et al. (1995)	$+$	$+$	[38]	$\pi^- p$
Aschenauer, Markushin (1996-97)	$+$	$+$	[39,40]	$\mu^- p,\, \mu^- d,\, \pi^- p$
Markushin (1999)	$+$	$+$	[34]	$\mu^- p,\, \mu^- d$
Terada, Hayano (1997)	$+$	$-$	[23]	$\pi^- p,\, \mu^- p,\, K^- p,\, \bar{p}p$
Faifman et al. (1999-2002)	$+$	$+$	[41,42]	$\mu^- p,\, K^- p,\, K^- d$
Koike (2002)	$+$	$+$	[43]	$\pi^- p,\, \mu^- p,\, K^- p,\, \bar{p}p$
Jensen, Markushin (2002)	$+$	$+$	[27,44–46]	$\pi^- p,\, \mu^- p,\, K^- p,\, \bar{p}p$

In the case of hadronic atoms, nuclear reactions can occur in S states where the x^- wave function has a significant overlap with the nuclear region. In antiprotonic hydrogen, the P state annihilation is also very important. In muonic, pionic, and kaonic atoms the cascade can be terminated by weak decay of the x^-.

The standard cascade model (SCM) for exotic hydrogen atoms developed in the 60–80th [16,20] was able to describe the basic features of the atomic cascade, and, with the use of a few phenomenological parameters, provided a fair qualitative description of the X-ray yields and absorption fractions [20,21,23]. For example, a scaling factor for the Stark mixing rates, so–called k_{STK}, [20] was often used in order to reproduce the X-ray yields in hadronic atoms. For the sake of simplicity, the SCM does not take into account the interplay between the internal and external degrees of freedom: the cascade rates calculated for a fixed value of the kinetic energy T are used in the kinetics calculation, with the value of T being treated as a fit parameter. The new generation of precision experiments, however, relies on a detailed understanding of the kinetic energy evolution during the cascade. The extended standard cascade model (ESCM) [33,34] was introduced in order to remove this limitation. The ESCM includes all the processes from the SCM and, in addition, the evolution of the energy distribution of the exotic atoms during the cascade is calculated from the master equation. A brief summary of the cascade models for exotic hydrogen atoms is given in Table 2.

A significant improvement over the previous calculations [33,34,38,40] was achieved in the recent cascade studies [27,44–47], where new results for the collisional processes were used. The cross sections for Stark and Coulomb transitions and elastic scattering were calculated simultaneously using the classical–trajectory Monte Carlo (CTMC) method for high n states [27,25]. The new results for Coulomb deexcitation replace the so–called chemical deexcitation which was often used in earlier cascade models [16,23] without any detailed understanding of the underlying dynamics. For the low n states, the exotic atom is described quantum mechanically and the target is treated as consisting of hydrogen atoms instead of molecules. The cross sections for Stark transitions, absorption during collision, and elastic scattering were calculated using a close–coupling model and the semiclassical (eikonal) approximation [24–26]. The cross sections for Auger transitions were calculated in the semiclassical approximation for the whole n range [27].

Using the new set of collisional cross sections one can describe the competition between the acceleration and deceleration mechanisms without employing any fit parameters. At present, the parameter–free cascade calculations are available for low–density targets where remaining theoretical uncertainties, which are mainly related to the Coulomb transitions and additional cascade mechanisms at the lower stage of the cascade, play a minor role (at low density the lower part of the atomic cascade is dominated by the radiative deexcitation), see Fig. 2. By confronting the theoretical predictions with experimental data one can determine the domain of validity of the ESCM and identify possible significant mechanisms beyond the ESCM.

3 Muonic Hydrogen

The hyperfine structure of the hydrogen–like atoms can be calculated with high precision in QED, see [48–50] and references therein. Further precision tests of QED in bound systems with ordinary hydrogen are presently limited by the experimental error in the proton charge radius r_p.

A precision measurement of the Lamb shift (the $2S - 2P$ energy difference) in muonic hydrogen can improve the uncertainty in the RMS proton charge radius r_p by more than one order of magnitude [6]. The proton radius is also a fundamental quantity of the baryon structure in QCD [51]. The QED calculations for the $\mu^- p$ Lamb shift are available at a precision level of 10^{-6} (see [52–54] and references therein):

$$\Delta E_{2S-2P} = 210.005(6) \text{ meV} - 5.166 r_p^2 \text{ meV} \cdot \text{fm}^2 \tag{1}$$

$$= 207.167(107) \text{ meV}, \quad r_p = 0.880(15) \text{ fm } [55]. \tag{2}$$

Since the finite size effect is about 2%, a precise measurement of the Lamb shift can provide a value of the proton radius with an accuracy of 0.1%. This goal is expected to be reached by the Muonic Hydrogen Lamb Shift Collaboration at PSI [6]. The HFS structure of the $n = 2$ states in muonic hydrogen is shown in Fig. 3a. While the $2P$ state quickly deexcites to the ground state $1S$, the $2S$ state, which is bound deeper than the $2P$ state, can be metastable, i.e. can have the lifetime significantly longer than the average cascade time. As shown in Fig. 3b, the idea of the Lamb shift measurement is to use a tunable laser to

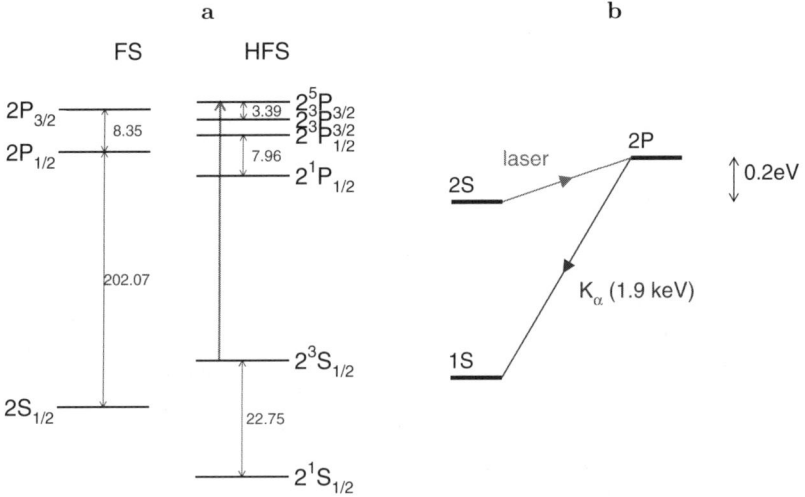

Fig. 3. a The energy structure of the $n = 2$ levels in muonic hydrogen (in units of meV, not to scale) [48].   **b** The laser induced transition $2S \to 2P$ followed by the radiative deexcitation $2P \to 1S$e

induce the $2S \to 2P$ transition followed by the X-ray transition $2P \to 1S$ that can be detected as a delayed K_α line.

The feasibility of the $\mu^- p$ Lamb shift experiment depends on the population and the life time of the metastable $2S$ state. The metastability of the $2S$ state depends, in particular, on its kinetic energy distribution. The $(\mu^- p)_{2S}$ atoms below the threshold of the $2S \to 2P$ Stark mixing, $T_{\mathrm{lab}} = 0.3$ eV, can be metastable because they are "immune" against the deexcitaion via the Stark transition to the $2P$ state followed by the fast radiative transition $2P \to 1S$ (the radiative quenching during the Stark collisions is small). For kinetic energies above the $2P$ threshold, the collisions with the target molecules lead to a deceleration competing with the depletion via the $2S \to 2P$ Stark transitions. The lifetime of the metastable $(\mu^- p)_{2S}$ is determined by the muon lifetime and the collisional (non-radiative) quenching via formation of molecular resonances [6,56].

The existence of long-lived metastable muonic hydrogen atoms in the $2S$ state has been demonstrated in a series of experiments at PSI [6,56] where the initial kinetic energy distributions of muonic hydrogen atoms in the *ground state* was measured using a time-of-flight (ToF) method. Low energy muons were stopped in a cylindrical low pressure gas target, and the time difference between the muon stop and the arrival of the $\mu^- p$ atom at the inner gold-coated target wall was measured. The target was located in a strong axial magnetic field which forced the muons to stop near the axis of the target, but did not affect the neutral $\mu^- p$ atom. The kinetic energy distributions were reconstructed from the ToF spectra in the density range $0.06 - 16$ mbar. The atomic cascade at 0.06 mbar is nearly pure radiative (the average number of collisions is less than one), therefore the kinetic energy distribution changes very little during the cascade. This allows

a

b

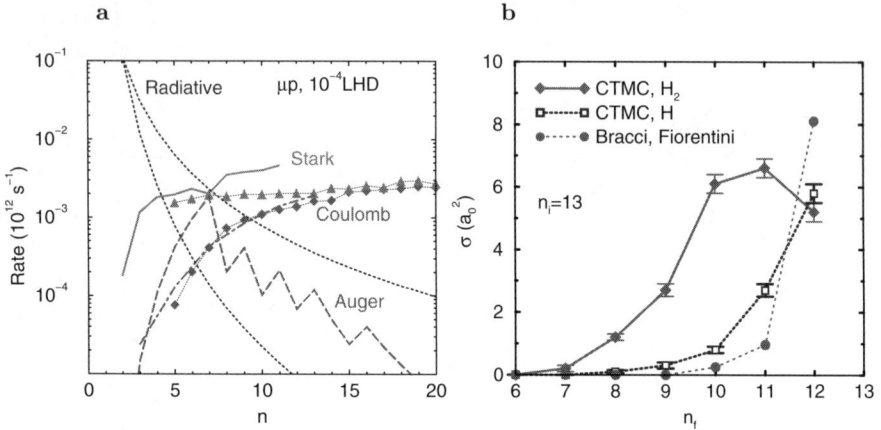

Fig. 4. a The l-average rates for muonic hydrogen with the lab kinetic energy $T = 1$ eV in gaseous target at 10^{-4} LHD (79 mbar). The lines are like in Fig. 2. The Coulomb deexcitation rate from [29] is shown with a dash-dotted line.   **b** The Coulomb deexcitation cross sections (in atomic units) for $(\mu p)_{n_i = 13}$ scattering from molecular and atomic hydrogen at the lab kinetic energy $T = 1$ eV as a function of the final state n_f [27]

one to determine the initial kinetic energy distribution in the very beginning of the atomic cascade and use it in the cascade calculations at higher densities.

In the recent theoretical studies of the atomic cascade in $\mu^- p$ [27], the evolution of the kinetic energy distribution has been calculated from the very beginning of the cascade. The cascade calculations used the new results for the collisional cross sections [26]. Figure 4a shows an example of the n-dependence of the rates of different cascade processes. Especially important was taking into account the molecular structure of hydrogen molecules in the Coulomb deexcitation of the $\mu^- p$ states with high n, see Fig. 4b. The Coulomb deexcitation leads to a significant acceleration in the very beginning of the atomic cascade due to the dominance of transitions with $\Delta n > 1$. This new version of ESCM provides a good description of many properties of the atomic cascade without using any fitting parameters. Figure 5a shows the calculated density dependence of the cumulative energy distribution in the ground state in comparison with the experimental data [56]. The calculated increase of the $(\mu^- p)_{1s}$ median kinetic energy with the density is in agreement with the data [56] for the initial conditions corresponding to kinetic energies about 0.5 eV and principal quantum number $n \approx 14$, see Fig. 5b. The molecular structure of the target is essential for explaining the observed density dependence of the kinetic energy distribution at the end of the cascade.

The calculated X-ray yields in muonic hydrogen are in good agreement with the data as shown in Fig. 6a. The measured ratio of the K_α and K_β yields in liquid hydrogen can be reproduced only if the metastable $2S$ state is depopulated

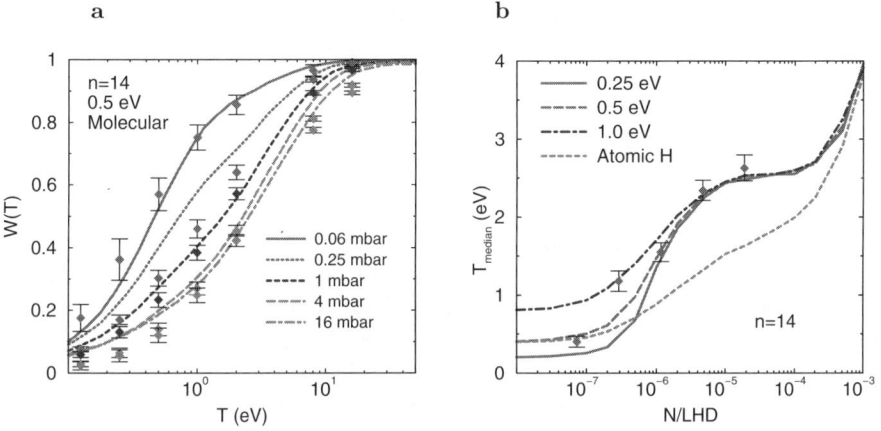

Fig. 5. **a** The calculated cumulative energy distribution $W(T)$ of the $\mu^- p$ atoms at the end of the cascade for initial conditions: $n_i = 14$ and $T_0 = 0.5$ eV [27]. The data are from [56].   **b** The calculated density dependence [27] of the median kinetic energy of the $\mu^- p$ at the end of the cascade for different initial average kinetic energies and $n_i = 14$. The result for atomic target is shown for $T_0 = 0.5$ eV. The data are from [56]

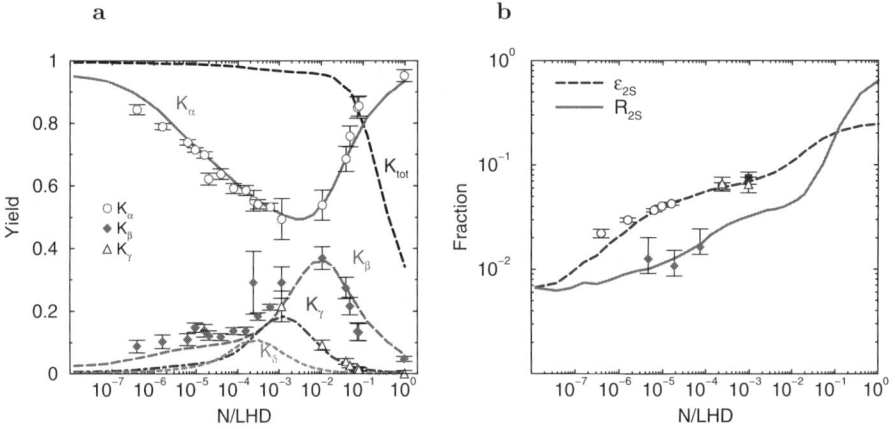

Fig. 6. **a** The density dependence of the relative X-ray yields, K_α, K_β, and K_γ, and the absolute total yield K_{tot} in muonic hydrogen [27]. The experimental data are from [57–59].   **b** The calculated $2S$ arrival probability ϵ_{2S} (*dashed line*) [27], in comparison with the values derived from X-ray measurements: [57] (*circles*), [60] (*triangles*), and [61] (*filled square*). The *solid line* shows the calculated metastable $2S$ fraction R_{2S}; the corresponding experimental data (*filled diamonds*) are from [56]

via a non-radiative quenching (see [56] and references therein for a discussion of the quenching mechanisms).

The calculated density dependence of the metastable $2S$ fraction is shown in Fig. 6b. The calculated arrival probability (the probability of the deexcitations from the states $n > 2$ to the $2S$ state) is in good agreement with the estimates based on the measured X-ray yields (see [56] and references therein). As discussed above, only a fraction of the initial $2S$ population can survive to form the metastable state if the radiative deexcitation of the $2P$ state is faster than the Stark mixing $2P \rightarrow 2S$. The relationship between these two processes changes at high density when the slowing down proceeds faster than the radiative transition $2P \rightarrow 1S$, and more than half of all muons in liquid hydrogen go to the $2S$ metastable state. However, the metastable $2S$ state in liquid hydrogen has a short lifetime because of the collisional quenching. Most suitable for the Lamb shift experiment is the density range around 1 mbar where the collisional quenching rate is comparable with the muon lifetime [6,56]. The calculated $2S$ metastable fraction at 1 mbar is about 1% in perfect agreement with the experimental data. In conclusion, the population and the lifetime of the metastable $(\mu^- p)_{2S}$ state have been found to be in the range suitable for a feasible measurement of the Lamb shift by means of laser induced spectroscopy [6].

4   Pionic Hydrogen

The pion-nucleon scattering lengths can be determined from the nuclear shift ΔE_{1S} and width Γ_{1S} of the $1S$ state of pionic hydrogen using the Deser type formula [1]

$$\Delta E_{1S} = -\frac{4\,a_{\pi^- p \rightarrow \pi^- p}}{r_B}\,E_{1S}(1 + \delta_E)\,, \tag{3}$$

$$\Gamma_{1S} = \frac{8q}{r_B}\left(1 + \frac{1}{P}\right)|a_{\pi^- p \rightarrow \pi^0 n}|^2\,E_{1S}(1 + \delta_\Gamma)\,, \tag{4}$$

where $a_{\pi^- p \rightarrow \pi^- p}$ is the $\pi^- p$ scattering length, $a_{\pi^- p \rightarrow \pi^0 n}$ is the S-wave amplitude of the charge exchange reaction $\pi^- p \rightarrow \pi^0 n$, $r_B = 222.56$ fm is the Bohr radius of pionic hydrogen, $E_{1S} = 3238$ eV is the point–like Coulomb binding energy of the $1S$ state, $P = 1.55$ is the Panofsky ratio taking into account the reaction channel $(\pi^- p) \rightarrow \gamma + n$, $q = 28.04$ MeV$/c$ is the CMS momentum of the π^0 in the charge exchange reaction, δ_E and δ_Γ are small electromagnetic correction. Precision πN scattering length measurements are of fundamental importance for the theory of strong interaction and the symmetry breaking in hadronic systems, see [7] and references therein. The goal of the new experiment at PSI [9] is to improve the present values [7]

$$\Delta E_{1S} = -7.108 \pm 0.013 \pm 0.034 \text{ eV}\,, \tag{5}$$

$$\Gamma_{1S} = 0.868 \pm 0.040 \pm 0.038 \text{ eV}\,, \tag{6}$$

by measuring the Γ_{1S} with an accuracy about 1%. This level of precision can be reached with an experimental setup where the $\pi^- p$ atoms are formed in a well defined volume using the PSI cyclotron trap, and the X–rays are measured using a bend–crystal spectrometer and a CCD detector [9,62]. Figure 7 shows an example of the measured K_β line in $\pi^- p$ in gaseous hydrogen at 3.5 bar together with the calibration line corresponding to the $6 \rightarrow 5$ transition in π^- ^{16}O. At

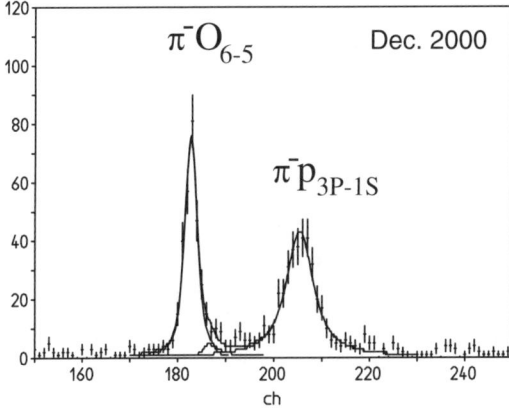

Fig. 7. The K_β line of the pionic hydrogen in gaseous target at a density equivalent to 3.5 bar [63]

this level of precision, the Doppler broadening corrections determine the ultimate limits of the $\pi^- p$ X-ray spectroscopy.

The atomic cascades in the $\mu^- p$ and $\pi^- p$ atoms are expected to have many similarities as they have comparable reduced masses. The main difference in the atomic cascade is due to the nuclear absorption in $\pi^- p$ that is especially important for low lying levels ($n < 6$) where the absorption significantly reduces the population of atomic states. In the upper part of the cascade, the absorption is less important because of the lower statistical weight of the nS states and smaller nuclear reaction rates:

$$\Gamma_{nS} = \frac{\Gamma_{1S}}{n^3} \ . \tag{7}$$

The $\pi^- p \rightarrow \pi^0 n$ reaction during the atomic cascade was studied using the neutron time–of–flight (n-ToF) method in [64–66] where the kinetic energy distribution at the instant of nuclear reaction was investigated by measuring the Doppler broadening of the n-ToF spectra. The discovery of a high energy component in the $\pi^- p$ kinetic energy distribution in [64] was the first direct evidence of the importance of the Coulomb–like acceleration in the atomic cascade. The kinetic energy distributions reconstructed from the n-ToF spectra in [66] reveal important details confirming the Coulomb acceleration mechanism, see Fig. 8.

The kinetic energy distribution at the instant of the radiative transitions $nP \rightarrow 1S$ is dependent on the initial state n and the target density, and so are the profiles of the X-ray lines. A typical example of the cumulative kinetic energy distribution and the corresponding Doppler broadening of the K_β line in a gaseous target is shown in Fig. 9 [44]. If the Doppler broadening corrections are applied properly, then the values for the nuclear width Γ_{1S} obtained from measurements at different target densities must agree with each other. This check

a   b

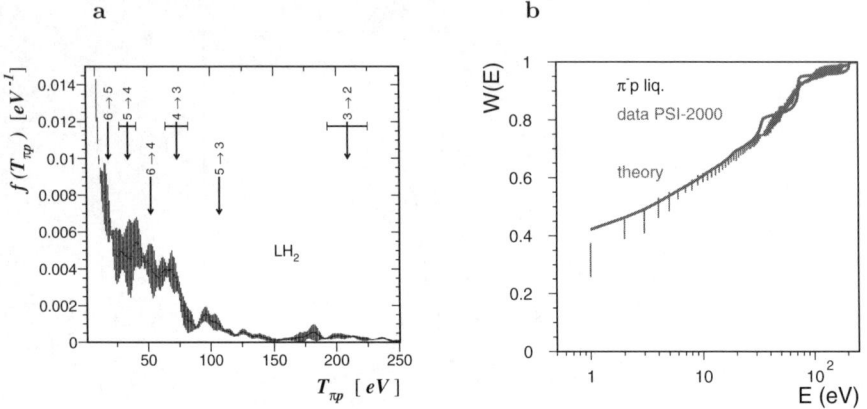

Fig. 8. a The energy distribution of the $\pi^- p$ atoms at the instant of the nuclear reaction $(\pi^- p) \to \pi^0 + n$ in liquid hydrogen reconstructed from the neutron ToF spectra [66]. The kinetic energies corresponding to the Coulomb transitions $n_i \to n_f$ are shown by arrows, with horizontal bars indicating the experimental energy resolution. **b** The calculated cumulative energy distribution of the $\pi^- p$ atoms at the instant of the nuclear reaction [44] in comparison with the experimental data [66]

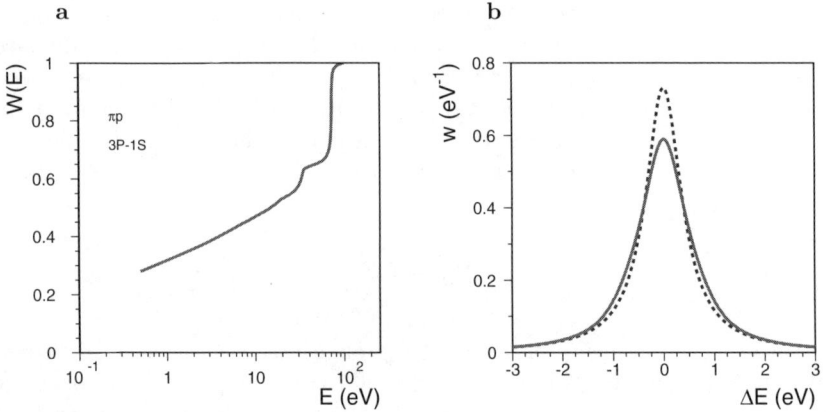

a   b

Fig. 9. a The calculated cumulative energy distribution of pionic hydrogen at the instant of the radiative transition $3P \to 1S$ at 3 bar. **b** The calculated profile of the K_β line in pionic hydrogen at 3 bar with the Doppler broadening taken into account (*solid line*) in comparison with the Lorentzian shape corresponding to the width of the $1S$ state (*dotted line*)

of the self–consistency of the data analysis is crucial for a reliable determination of systematic errors.

Additional checks of the cascade model, which is used in the calculations of the Doppler broadening corrections, can be done by using the same model for the muonic hydrogen and comparing the calculated X-ray profiles with meaa

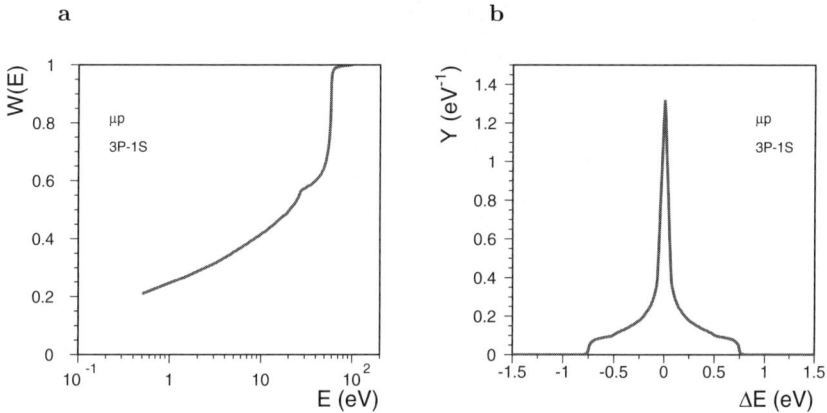

b

Fig. 10. a The calculated cumulative energy distribution of muonic hydrogen at the instant of the radiative transition $3P \rightarrow 1S$ at 3 bar.   **b** The calculated Doppler broadening of the K_β line in muonic hydrogen at 3 bar.

sured ones. Figure 10 demonstrates the cumulative energy distribution for the $\mu^- p$ atom at the instant of the radiative transition $3P \rightarrow 1S$ at 3 bar and the corresponding Doppler profile of the K_β line. There is a clear similarity between the $\mu^- p$ and $\pi^- p$ energy distributions at the same experimental conditions (compare Figs. 9a and 10a) that warrant a detailed study of the Doppler broadening of the $\mu^- p$ X-ray lines. The measurements of the $\mu^- p$ X-ray Doppler broadening are planned as a part of the pionic hydrogen experiment [9].

5   Kaonic Hydrogen

Like in the $\pi^- p$ case, the kaon-nucleon scattering lengths can be obtained by measuring the shift and width of the K X-ray lines in $K^- p$. Around 1980 three measurements of X-rays in liquid hydrogen [67–69] were carried out. The experiments suffered from poor statistics and large background but all three measurements found peaks in the X-ray spectrum which were attributed to the K X-ray lines in $K^- p$. The resulting *positive* $1S$ energy shift contradicted the results based on analyses of scattering data which led to a *negative* $1S$ energy shift. This discrepancy, known as "the kaonic hydrogen puzzle", was resolved when the first (and only till now) reliable measurement was done at KEK [70] using a cryogenic gas target. The shift and width were determined to be [70] $\Delta E_{1S} = 323 \pm 63(\text{stat}) \pm 11(\text{sys})$ eV and $\Gamma_{1S} = 407 \pm 208(\text{stat}) \pm 100(\text{sys})$ eV consistent with the scattering data. The energy shifts reported in the earlier measurements were thus artifacts due to the poor quality of the spectra. The DEAR experiment [11], currently in progress at Frascati, aims at determining the $K^- p$ (as well as the $K^- d$) shift and width with a precision of 1%.

The kaonic hydrogen case differs from the pionic one in two major respects: first, the $1S$ width is so large that Doppler corrections are completely negligible.

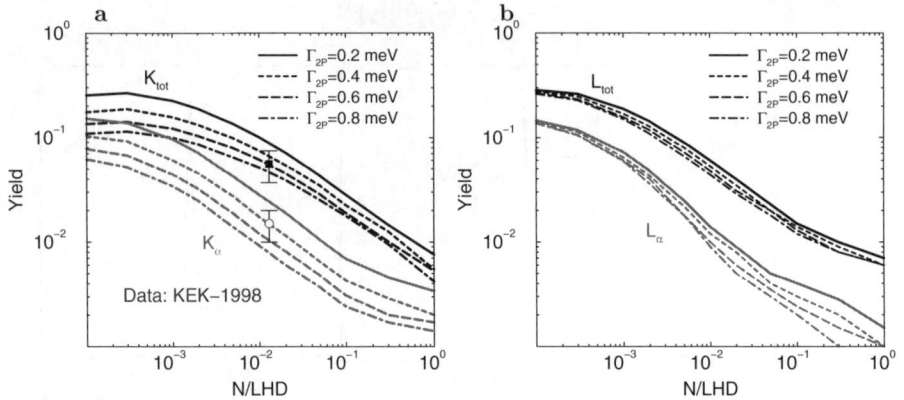

Fig. 11. a The density dependence of the X-ray yields, K_{tot} and K_α, in kaonic hydrogen for different values of $2P$ hadronic width. The experimental data are from [10]. **b** The density dependence of the X-ray yields, L_{tot} and L_α, in kaonic hydrogen

The very challenging problem of calculating the kinetic energy distribution at the instant of the radiative transition does not occur here. Secondly, because of the large strong interaction width, the K X-ray lines overlap and form the so-called K line complex which must be disentangled in the data analysis [71].

The kaonic hydrogen experiments can benefit from a detailed understanding of the atomic cascade. For example, in order to determine the optimal experimental conditions for achieving the highest count rates, reliable theoretical predictions of the absolute X-ray yields are needed. In addition, predictions of relative K yields are useful for the data analysis of the K X-ray spectrum as a cascade constrained fit can lead to an improved determination of the $1S$ width [10].

The predictions of the ESCM depend on the strong interaction $1S$ shift and width ΔE_{1S} and Γ_{1S} and the poorly known $2P$ strong interaction width: Γ_{2P}^{had}. The widths of the higher lying nP states are given by

$$\Gamma_{np}^{\text{had}} = \frac{32(n^2 - 1)}{3n^5} \Gamma_{2p}^{\text{had}} . \tag{8}$$

In the following the $1S$ shift and width are fixed at the central values of the KEK result and Γ_{2P}^{had} is treated as a free parameter[1].

Figure 11a shows the absolute K yields calculated for different values of the $2P$ width. The total yield decreases with increasing density from 10-25% at 10^{-3} LHD to 0.4-0.8% at LHD. This behavior is due to the Stark effect: with increasing density the K^-p atoms undergo more collisions resulting in a more efficient feeding of the nS and nP states from which the nuclear reactions take place. Figure 11b shows the calculated L_{tot} and L_α yields. Compared to the K yields there is a much weaker dependence on the hadronic $2P$ width.

---

[1] One effective width for the $2P$ state corresponds to the four partial wave amplitudes $P_{I,2J} = P_{01}, P_{03}, P_{11}, P_{13}$ in KN scattering.

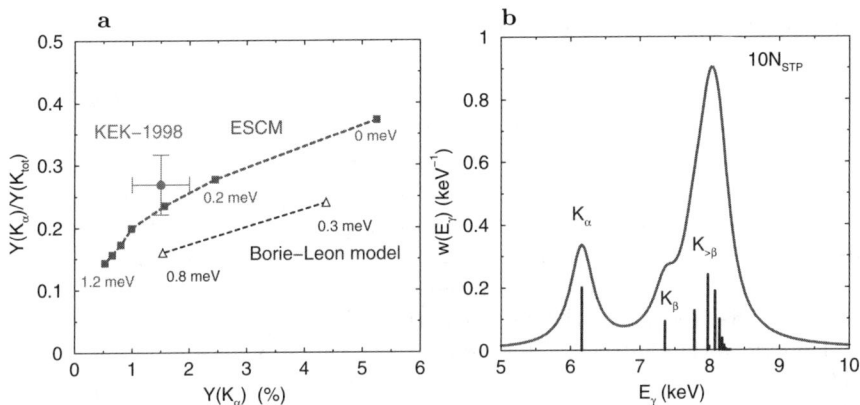

Fig. 12. a The correlation between the relative and absolute K_α yields in kaonic hydrogen at 10 bar. The results of the ESCM cascade calculations are shown in comparison with the experimental data from KEK [10]. The hadronic width of the $2P$ state varies between 0 and 1.2 meV in steps of 0.2 meV. BL is the result reported in [10] of the Borie–Leon model with $\Gamma_{2P}^{\mathrm{had}} = 0.3$ meV and 0.8 meV.   b The K line profile in kaonic hydrogen at 10 bar for $\Gamma_{2P}^{\mathrm{had}}=0.5$ meV

Though L X-rays in kaonic hydrogen were observed at KEK [10], the yields could not be obtained because of poor knowledge of the detector efficiency in the relevant energy region. A measurement of the intensity ratio $Y(K_\alpha)/Y(L_{\mathrm{tot}})$ would, however, be very useful because this ratio allows one to obtain the $2P$ width with high accuracy (like it was done in $\bar{p}p$).

Figure 12a shows the relative K_α yield, $Y(K_\alpha)/Y(K_{\mathrm{tot}})$, vs. the absolute K_α yield. In the Borie–Leon model with the tuning parameters $k_{\mathrm{STK}}=1.8$ and $T = 1.0$ eV used by [10], it was not possible to get agreement with the data by varying $\Gamma_{2P}^{\mathrm{had}}$. In order to explain the measured relative K_α yield, a hadronic width about 0.3 meV was needed but this led to an absolute K_α yield of 4.38% which is significantly above the experimental result. The absolute K_α yield can be reproduced with the width 0.8 meV, but then the calculated relative yield becomes inconsistent with the data. The situation improves greatly in the ESCM, as agreement can be reached for a range of values for the hadronic width: the best agreement for the width is found for $\Gamma_{2P}^{\mathrm{had}} = 0.2 - 0.6$ meV.

The K line profile at 10 bar is shown in Fig. 12b. The broadening due to the strong interaction makes the disentangling of the K line complex a difficult problem [71]: only the K_α peak is clearly separable from the rest of the K lines. The relative intensities of the K lines can be qualitatively understood as follows: at 10 bar the strong absorption from the S and P states is important for $n \leq 10$. The strong absorption from the excited states prevents most of the K^-p atoms from reaching the lower levels and as a result, the K_δ yield is significantly higher than the K_γ and the K_β yields.

6   Antiprotonic Hydrogen

A series of spectroscopy experiments at the Low-Energy Antiproton Ring (LEAR) at CERN has lead to a determination of $1S$ and $2P$ shifts and widths in antiprotonic hydrogen (see [13] and references therein). From the point of view of cascade calculation the $2P$ state poses a challenging problem because of potentially large Doppler corrections to the measured width of the $3D \to 2P$ radiative transition from which the strong interaction width is obtained [72]. As in the case of pionic hydrogen a reliably cascade model is needed to calculate the kinetic energy distribution at the instant of the radiative transition. Presently, the ESCM provides the best choice for dealing with this problem.

The cascade in antiprotonic hydrogen is expected to start with a principal quantum number $n \approx 30$. The collisional rates for the kinetic energy 1 eV and the density 10^{-4} LHD are shown in Fig. 13a. Like for the other exotic hydrogen atoms, the dominant collisional deexcitation mechanism at high n is the Coulomb deexcitation which is much stronger than the external Auger effect. A non-thermal kinetic energy distribution is therefore expected to develop during the upper part of the cascade similar to the $\mu^- p$ case. An important feature of the initial stage of the atomic cascade is that the Coulomb deexcitation proceeds via large jumps in n; this has significant effects in the cascade times and kinetic energy distributions. A typical example of the dependence of the Coulomb deexcitation cross sections on the final state n_f is shown in Fig. 13b for the initial states $n_i = 20, 25, 30$ and the kinetic energy $T = 1$ eV. The cross sections have been calculated in the classical-trajectory Monte Carlo model on molecular hydrogen target [27].

The cascade time measurements at low density done by the OBELIX collaboration [73] provide important confirmation of the domination of collisional

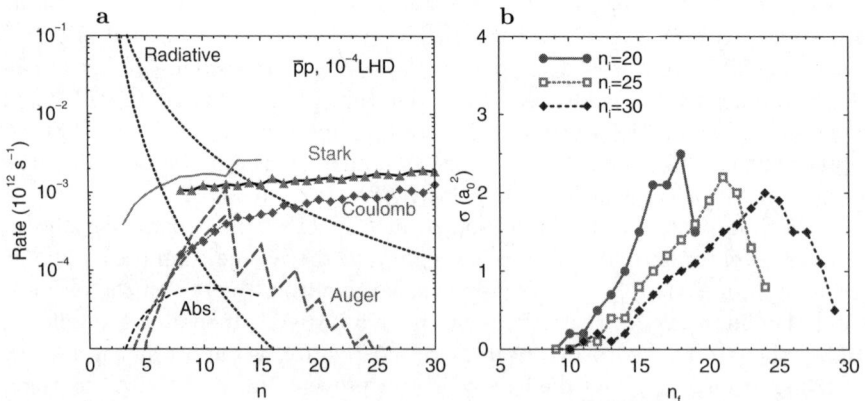

Fig. 13. a The n dependence of the rates in antiprotonic hydrogen at 10^{-4} LHD and $T = 1$ eV. The lines are as in Fig. 2. In addition the rate for nuclear absorption during collisions is shown with a *dashed curve*. **b** The n_f dependence of the Coulomb cross sections for $n_i = 20, 25, 30$. The lab kinetic energy is 1 eV

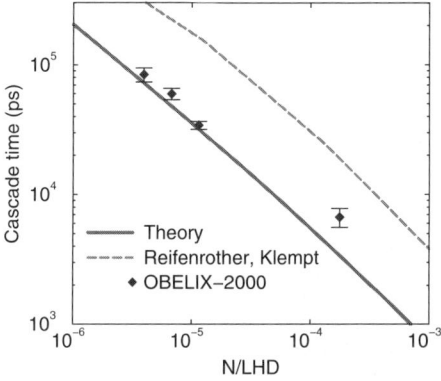

Fig. 14. The density dependence of the cascade time in antiprotonic hydrogen calculated in the ESCM [27] in comparison with the earlier calculations [35] and the experimental data [73]

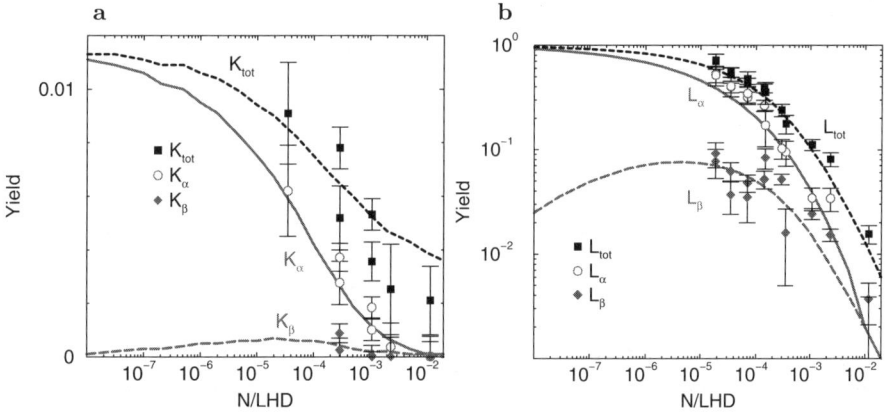

Fig. 15. The density dependence of the X-ray yields in antiprotonic hydrogen [27]. **a** The K yields.   **b** The L yields. The experimental data are from [74–77]

deexcitation with large Δn at high n. The density dependence of the cascade time is shown in Fig. 14. There is good agreement between the result of the ESCM and the experimental data at low densities ($< 10^{-5}$ LHD) whereas the earlier cascade model [35] predicted a cascade time that was about a factor of five larger. The experimental data support the classical Monte Carlo result of fast collisional deexcitation in the upper part of the cascade.

The K and L X-ray yields in $\bar{p}p$ calculated in the ESCM [27] are in good agreement with the experimental data [74–77] as shown in Fig. 15. The calculated yields were obtained without using any phenomenological tuning parameters that were necessary in earlier cascade calculations [20]. Typical examples of fits in the Borie-Leon model are $k_{\mathrm{STK}} = 2.16^{+0.23}_{-0.20}$, $T = 1.0^{+0.3}_{-0.2}$ eV [76] and $k_{\mathrm{STK}} = 1.94 \pm 0.13$, $T = 1.0 \pm 0.2$ eV [77].

The shifts of the $2P$ states are caused by vacuum polarization, electromagnetic fine and hyperfine splitting, and hadronic effects. The electromagnetic splitting of the $2P$ level [72,13] forms two groups of hyperfine states with the $^3P_0$

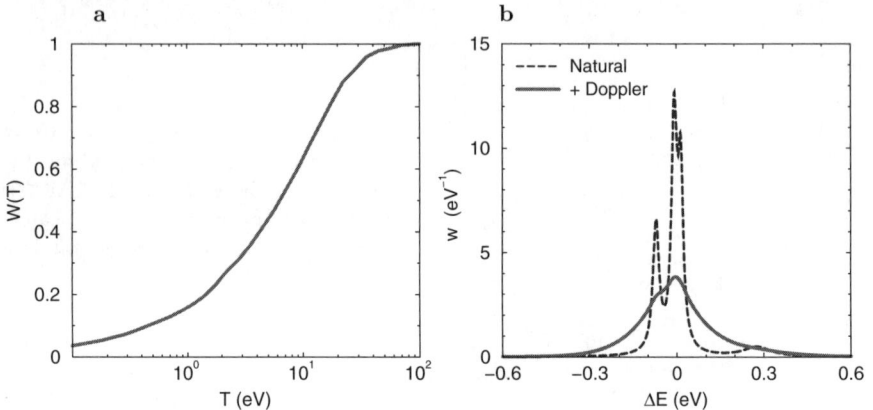

Fig. 16. **a** The cumulative kinetic energy distribution at the instant of absorption in antiprotonic hydrogen at 22 mbar.   **b** The theoretical L_α line profile at 22 mbar with and without Doppler broadening

state separated about 200 meV from the close–lying components $^3P_2$, $^3P_1$, $^1P_1$. The measurements of the $3D \to 2P$ line shape [72] have made it possible to determine the additional hadronic shifts and widths and test different models for the $\bar{p}p$ interaction.

The Coulomb deexcitation in the upper part of the cascade leads to acceleration of the $\bar{p}p$ atoms to energies of several electron-Volts. The cumulative kinetic energy distribution at 22 mbar at the instant of nuclear absorption is shown in Fig. 16a, with the median energy being $T_m \sim 6.5$ eV. The predicted energy distribution leads to a significant Doppler broadening of the L X-ray lines shown in Fig. 16b. The Doppler broadening of the $3D \to 2P$ line is comparable with the hardonic widths of the $2P$ states, and, therefore, it must be taken into account in the analysis of the X-ray spectra. In the earlier data analysis [72], it was assumed that the $\bar{p}p$ atoms were thermalized due to elastic and Auger collisions.

7   Conclusion

A reliable theoretical model of the atomic cascade is essential for the current generation of precision experiments with light exotic atoms. One of the main problems is a detailed description of the kinetic energy distributions at the instant of the radiative transitions in pionic and antiprotonic hydrogen. The corresponding Doppler broadening corrections determine the ultimate precision with which the strong interaction widths can be determined from the X-ray spectra. The precision experiments with exotic atoms are typically low–count–rate measurements, therefore they can significantly benefit from theoretical guidance in finding optimum experimental conditions to minimize statistical errors. In this situation, it is not sufficient to rely any more on the phenomenological cascade models that can be fit to the data with some *ad hoc* parameters. To meet these experimental challenges one needs the *ab initio* cascade calculations.

The first systematic implementation of this approach was undertaken in the recent calculations of the atomic cascades in $\mu^- p$, $\pi^- p$, $K^- p$, and $\bar{p}p$ in the framework of the extended standard cascade model [27,44]. For the first time, the acceleration and deceleration mechanism have been taken into account from the very beginning of the atomic cascade, and the evolution of the kinetic energy distribution has been calculated in a straightforward way from the master equation. As a result, a number of long standing problems have been solved, in particular, a long standing puzzle of chemical deexcitation [16], which was included in many cascade calculations in a pure phenomenological way for about 40 years. The underlying dynamics actually corresponds to the Coulomb deexcitation of highly excited states in collisions with hydrogen molecules, with the target molecular structure being important for large decrease of the principal quantum number n in a single collision. The consequences of this mechanism can be directly seen in the energy distributions of muonic hydrogen in the ground state at low densities [6,56]. The ESCM also provides a good description of the data on the cascade times and the X-ray yields in antiprotonic hydrogen, with which the earlier calculations always had problems.

The ESCM will be subjected to stringent tests in the precision experiments with pionic and muonic hydrogen. The Doppler broadening of the $\mu^- p$ X-ray lines will be measured by the pionic hydrogen collaboration as a cross check of the atomic cascade model because the acceleration and deceleration processes are expected to be similar in the $\pi^- p$ and $\mu^- p$ atoms. Another critical test of the cascade calculations will come from the determination of the strong interaction width Γ_{1S} of the $(\pi^- p)_{1S}$ state from the radiative transitions $nP \rightarrow 1S$ at different densities. While the X-ray line shapes depend, because of the Doppler broadening, on the initial nP state and the target density, the corresponding values of Γ_{1S} obtained by applying theoretically calculated corrections must agree with each other.

The present version of the ESCM is definitely not the final one. In particular, there are indications that additional mechanisms (presently not included in the calculations), like the resonant formation of molecular states [78], are important at the lower stage of the cascade. These mechanisms are not very important at low density, when the lower part of the cascade is dominated by the radiative deexcitation, but may not be neglected at high density. Finding disagreements between the "no–free–parameters" cascade calculations and experimental data, in particular, the dependence of these effects on the experimental conditions, would be important for identifying these additional cascade mechanisms. Thus, the precision measurements with exotic atoms are mutually beneficial for both nuclear and atomic physics.

Acknowledgment

The authors thank M. Daum, D. Gotta, P. Hauser, H.J. Leisi, F. Kottmann, C. Petitjean, R. Pohl, L.M. Simons, and D. Taqqu for fruitful collaboration and numerous discussions on the problems of atomic cascade.

References

1. S. Deser et al.: Phys. Rev. **96**, 774 (1954)
2. V.S. Popov *et al*: Sov. Phys. JETP **53**, 650 (1981)
3. J. Carbonell and K. V. Protasov: J. Phys. **G18**, 1863 (1992)
4. V.E. Lubovitskij, A. Rusetsky: Phys. Lett. B **494**, 9 (2000)
5. D. Taqqu et al.: Hyperf. Interact. **119**, 311 (1999)
6. R. Pohl et al.: in *Hydrogen Atom: Precision Physics of Simple Atomic Systems*, Eds. S.G. Karshenboim et al., Lecture Notes in Physics **570**, Springer, Berlin (2001), p. 454
7. H.C. Schröder et al.: Eur. Phys. J. C **21**, 473 (2001)
8. D. Gotta: πN Newsletter **15**, 276 (1999)
9. D.F. Anagnostopoulos et al.: in *Hydrogen Atom: Precision Physics of Simple Atomic Systems*, Eds. S.G. Karshenboim et al., Lecture Notes in Physics **570**, Springer, Berlin (2001), p. 508
10. T.M. Ito et al.: Phys. Rev. C **58**, 2366 (1998)
11. The DEAR Collaboration, R. Baldini et al.: DAΦNE Exotic Atom Research, Preprint LNF-95/055 (1995)
12. M. Augsburger et al.: Nucl. Phys. A **658**, 149 (1999)
13. D.F. Anagnostopoulos et al.: in *Hydrogen Atom: Precision Physics of Simple Atomic Systems*, Eds. S G. Karshenboim et al., Lecture Notes in Physics **570**, Springer, Berlin (2001), p. 489
14. H. Anderhub et al.: Phys. Lett. B **101**, 151 (1981)
15. L.M. Simons: Phys. Scripta T **22**, 90 (1988)
16. M. Leon and H.A. Bethe: Phys. Rev. **127**, 636 (1962)
17. J.S. Cohen: Phys. Rev. A **59**, 1160 (1999)
18. G.Ya. Korenman: Hyperf. Interact. **101/102**, 81 (1996)
19. H.A. Bethe and E.E. Salpeter: *Quantum Mechanics of One- and Two-Electron Atoms* (Academic Press, New York, 1957)
20. E. Borie and M. Leon: Phys. Rev. A **21**, 1460 (1980)
21. V.E. Markushin: Sov. Phys. JETP **53**, 16 (1981)
22. V.P. Popov and V.N. Pomerantsev: Hyperf. Interact. **101/102**, 133 (1996); Hyperf. Interact. **119**, 133 (1999)
23. T.P. Terada and R.S. Hayano: Phys. Rev. C **55**, 73 (1997)
24. T.S. Jensen and V.E. Markushin: Nucl. Phys. A **689**, 537 (2001)
25. T.S. Jensen and V.E. Markushin: Proceedings of μCF01, Hyperf. Interact. (2002) (in press)
26. T.S. Jensen and V.E. Markushin: Eur. Phys. J. D, **19**, 157 (2002)
27. T.S. Jensen and V.E. Markushin: Eur. Phys. J. D (2002) (in press); e-Print Archive: physics/0205076, physics/0205077
28. A.P. Bukhvostov and N.P. Popov: Sov. Phys. JETP **55**, 12 (1982)
29. L. Bracci and G. Fiorentini: Nuovo Cim. A **43**, 9 (1978)
30. L.I. Ponomarev and E.A. Solov'ev: Hyperf. Interact. **119**, 55 (1999)

31. V.P. Popov and V.N. Pomerantsev: Hyperf. Interact. **119**, 137 (1999)
32. V.V. Gusev et al.: Hyperf. Interact. **119**, 141 (1999)
33. V.E. Markushin: Phys. Rev. A **50**, 1137 (1994)
34. V.E. Markushin: Hyperf. Interact. **119**, 11 (1999)
35. G. Reifenröther and E. Klempt: Nucl. Phys. A **503**, 885 (1989)
36. G. Reifenröther and E. Klempt: Phys. Lett. B **248**, 250 (1990)
37. W. Czaplinski et al.: Phys. Rev. A **50**, 518 (1994)
38. E.C. Aschenauer et al.: Phys. Rev. A **51**, 1965 (1995)
39. E.C. Aschenauer and V.E. Markushin: Hyperf. Interact. **101/102**, 97 (1996)
40. E.C. Aschenauer and V.E. Markushin: Z. Phys. D **39**, 165 (1997)
41. M.P. Faifman et al.: in *Frascati Phys. Series* **XVI**, 637 (1999)
42. M.P. Faifman and L.I. Men'shikov: Proceedings of μCF01, Hyperf. Interact. (2002) (in press)
43. T. Koike: Proceedings of μCF01, Hyperf. Interact. (2002) (in press)
44. V.E. Markushin and T.S. Jensen: Proceedings of μCF01, Hyperf. Interact. (2002) (in press)
45. T.S. Jensen: Diss. Zürich Univ., Zürich (2002)
46. T.S. Jensen and V.E. Markushin: πN Newsletter **16**, 358 (2002)
47. V.E. Markushin and T.S. Jensen: Nucl. Phys. A **691**, 318 (2001)
48. K. Pachucki: Phys. Rev. A **53**, 2092 (1996)
49. K. Jungmann et al.: in *Hydrogen Atom: Precision Physics of Simple Atomic Systems*, Eds. S.G. Karshenboim et al., Lecture Notes in Physics **570**, Springer, Berlin (2001), p. 446
50. S.G. Karshenboim and V.G. Ivanov: Eur. J. Phys. D **19**, 13 (2002)
51. J.L. Friar: *this volume*, (2002)
52. T. Kinoshita and M. Nio: Phys. Rev. Lett. **82**, 3240 (1999)
53. K. Pachucki: Phys. Rev. A **60**, 3593 (1999)
54. J.L. Friar et al.: Phys. Rev. A **59**, 4061 (1999)
55. R. Rosenfelder: Phys. Lett. B **479**, 318 (2000)
56. R. Pohl: Diss. ETH No. 14096, Zürich (2001)
57. H. Anderhub et al.: Phys. Lett. B **143**, 65 (1984)
58. M. Bregant et al.: Phys. Lett. A **241**, 344 (1998)
59. B. Lauss et al.: Phys. Rev. Lett. **80**, 3041 (1998)
60. A. Anderhub et al.: Phys. Lett. B **71**, 443 (1977)
61. P.O. Egan et al.: Phys. Rev. A **23**, 1152 (1981)
62. PSI Experiment R-98-01. (http://pihydrogen.web.psi.ch/)
63. D.F. Anagnostopoulos et al.: in PSI Scientific Rep. 2000, Vol. 1, p. 14
64. J.E. Crawford et al.: Phys. Rev. D **43** 46 (1991)
65. A. Badertscher et al.: Phys. Lett. B **392** 278 (1997)
66. A. Badertscher et al.: Europhys. Lett. **54**, 313 (2001)
67. J.D. Davies et al.: Phys. Lett. B **83**, 55 (1979)
68. M. Izycki et al.: Z. Phys. A **297**, 11 (1980)
69. P.M. Bird et al.: Nucl. Phys. A **404**, 482 (1983)
70. M. Iwasaki et al.: Phys. Rev. Lett. **78**, 3067 (1997)
71. C. Guaraldo et al.: Nuovo Cim. A **110**, 1347 (1997)
72. D. Gotta et al.: Nucl. Phys. A **660**, 283 (1999)
73. A. Bianconi et al.: Phys. Lett. B **487**, 224 (2000)
74. C. A. Baker et al.: Nucl. Phys. A **483**, 631 (1988)
75. C. W. E. Eijk et al.: Nucl. Phys. A **486**, 604 (1988)
76. R. Bacher et al.: Z. Phys. A **334**, 93 (1989)
77. K. Heitlinger et al.: Z. Phys. A **342**, 359 (1992)
78. S. Jonsell, J. Wallenius and P. Froelich: Phys. Rev. A **59**, 3440 (1999)

The Structure of Light Nuclei and Its Effect on Precise Atomic Measurements

James L. Friar

Theoretical Division, Los Alamos National Laboratory, Los Alamos, NM, USA
friar@lanl.gov

Abstract. This review consists of three parts: (a) what every atomic physicist needs to know about the physics of light nuclei; (b) what nuclear physicists can do for atomic physics; (c) what atomic physicists can do for nuclear physics. A brief qualitative overview of the nuclear force and calculational techniques for light nuclei will be presented, with an emphasis on debunking myths and on recent progress in the field. Nuclear quantities that affect precise atomic measurements will be discussed, together with their current theoretical and experimental status. The final topic will be a discussion of those atomic measurements that would be useful to nuclear physics, and nuclear calculations that would improve our understanding of existing atomic data.

1   Introduction

....numerical precision is the very soul of science.....

This quote [1] from Sir D'Arcy Wentworth Thompson, considered by many to be the first biomathematician, could well serve as the motto of the field of precise atomic measurements, since precision is the *raison d'être* of this discipline. I have always been in awe of the number of digits of accuracy achievable by atomic physics in the analysis of simple atomic systems [2]. Nuclear physics, which is my primary field and interest, must usually struggle to achieve three digits of numerical significance, a level that atomic physics would consider a poor initial effort, much less a decent final result.

The reason for the differing levels of accuracy is well known: the theory of atoms is QED, which allows one to calculate properties of few-electron systems to many significant figures [3]. On the other hand, no aspect of nuclear physics is known to that precision. For example, a significant part of the "fundamental" nuclear force between two nucleons must be determined phenomenologically by utilizing experimental information from nucleon-nucleon scattering [4], very little of which is known to better than 1%. In contrast to that level of precision, energy-level spacings in few-electron atoms can be measured so precisely that nuclear properties influence significant digits in those energies [5]. Thus these experiments can be interpreted as either a measurement of those nuclear properties, or corrections must be applied to eliminate the nuclear effects so that the resulting measurement tests or measures non-nuclear properties. That is the purview of this review.

The single most difficult aspect of a calculation for any theorist is assigning uncertainties to the results. This is not always necessary, but in calculating nuclear corrections to atomic properties it is essential to make an effort. That is just another way to answer the question, "What confidence do we have in our results?" Because it is important for atomic physicists to be able to judge nuclear results to some degree, this discussion has been slanted towards answers to two questions that should be asked by every atomic physicist. The first is: "What confidence should I have in the values of nuclear quantities that are required to analyze precise atomic experiments?" The second question is: "What confidence should I have that the nuclear output of my atomic experiment will be put to good use by nuclear physicists?"

2   Myths of Nuclear Physics

Every field has a collection of myths, most of them being at least partially true at one time. Myths propagate in time and distort the reality of the present. A number of these are collected below, some of which the author once believed. The resolution of these "beliefs" also serves as a counterpoint to the very substantial progress made in light-nuclear physics in the past 15 years, which continues unabated.

My myth collection includes:

• The strong interactions (and consequently the nuclear force) aren't well understood, and nuclear calculations are therefore unreliable.

• Large strong-interaction coupling constants mean that perturbation theory doesn't converge, implying that there are no controlled expansions in nuclear physics.

• The nuclear force has no fundamental basis, implying that calculations are not trustworthy.

• You cannot solve the Schrödinger equation accurately because of the complexity of the nuclear force.

• Nuclear physics requires a relativistic treatment, rendering a difficult problem nearly intractable.

All of these myths had some (even considerable) truth in the past, but today they are significant distortions of our current level of knowledge.

3   The Nuclear Force

Most of the recent progress in understanding the nuclear force is based on a symmetry of QCD, which is believed to be the underlying theory of the strong interactions (or an excellent approximation to it). It is generally the case that our understanding of any branch of physics is based on a framework of symmetry principles. QCD has "natural" degrees of freedom (quarks and gluons) in terms of which the theory has a simple representation. The (strong) chiral symmetry of QCD results when the quark masses vanish, and is a more complicated analogue of the chiral symmetry that results in QED when the electron mass vanishes. The latter symmetry explains, for example, why (massless or high-energy) electron scattering from a spherical (i.e., spinless) nucleus vanishes in the backward direction.

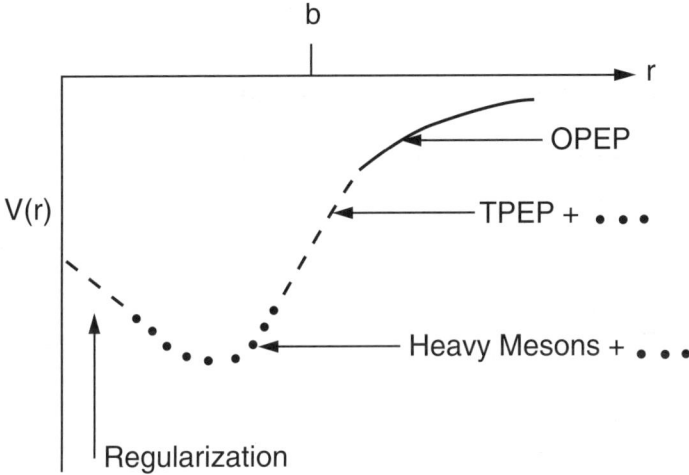

Fig. 1. Cartoon of the nuclear potential, $V(r)$, showing regions of importance

The problem with this attractive picture is that it does not involve the degrees of freedom most relevant to experiments in nuclear physics: nucleons and pions. It is nevertheless possible to "map" QCD (expressed in terms of quarks and gluons) into an "equivalent" or surrogate theory expressed in terms of nucleon and pion degrees of freedom. This surrogate works effectively only at low energy. The small-quark-mass symmetry limit becomes a small-pion-mass symmetry limit. In general this (slightly) broken-symmetry theory has $m_\pi c^2 \ll \Lambda$, where the pion mass is $m_\pi c^2 \cong 140$ MeV and $\Lambda \sim 1$ GeV is the mass scale of QCD bound states (heavy mesons, nucleon resonances, etc.). The seminal work on this surrogate theory, now called chiral perturbation theory (or χPT), was performed by Steve Weinberg [6], and many applications to nuclear physics were pioneered by his student, Bira van Kolck [7]. From my perspective they demonstrated two things that made an immediate impact on my understanding of nuclear physics [8]: (1) There is an alternative to perturbation theory in coupling constants, called "power counting," that converges geometrically like $(Q/\Lambda)^N$, where $Q \sim m_\pi c^2$ is a relevant nuclear energy scale, and the exponent N is constrained to have $N \geq 0$; (2) nuclear physics mechanisms are severely constrained by the chiral symmetry. These results provide nuclear physics with a well-founded rationale for calculation.

This scheme divides the nuclear-force regime in a natural way into a long-range part (which implies a low energy, Q, for the nucleons) and a short-range part (corresponding to high energy, Q, between nucleons). This is indicated in Fig. 1, which is a cartoon of the potential between two nucleons meant only to indicate significant regions and mechanisms. Since χPT is effective only at low energies, we expect that only the long-range part of the nuclear force can be treated successfully by utilizing only the pion degrees of freedom. This would be the region with $r > b$. We need to resort to phenomenology (i.e., fitting to

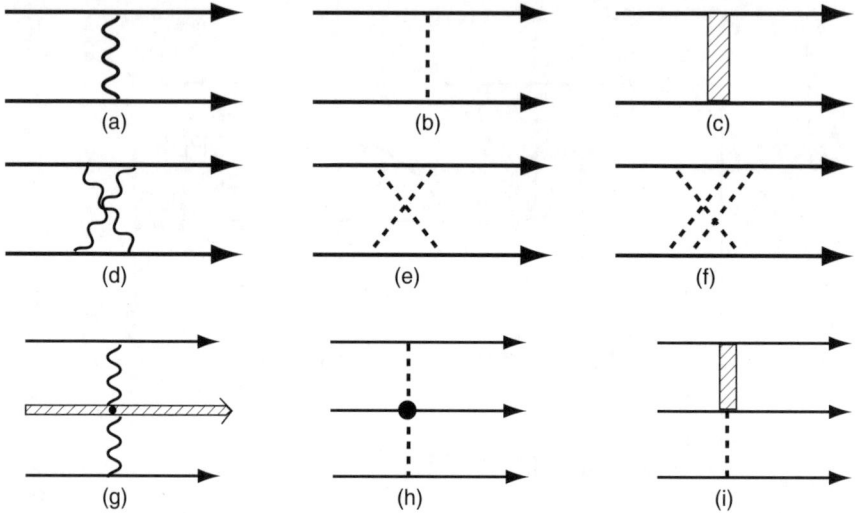

Fig. 2. First- and second-order (in α, the fine structure constant) atomic interactions resulting from photon exchange are shown in the left-most column, where solid lines are electrons, wiggly lines are photons, and the shaded line is a nucleus. The analogous nuclear interactions resulting from pion exchange are shown in the middle column, where solid lines are nucleons and dashed lines are pions. Nuclear processes involving short-range interactions (shaded vertical areas) are shown in the right-most column, together with a three-pion-exchange interaction

nucleon-nucleon scattering data) to treat systematically the short-range part of the interaction ($r < b$).

The long-range nuclear force is calculated in much the same way that atomic physics calculates the interactions in an atom using QED. Both are illustrated in Fig. 2. The dominant interaction between two nucleons is the exchange of a single pion illustrated in Fig. 2b (One-Pion-Exchange Potential or "OPEP") and denoted V_π. Its atomic analogue is one-photon exchange in Fig. 2a (containing the dominant Coulomb force). Because it is such an important part of the nuclear potential, it is fair to call V_π the "Coulomb force" of nuclear physics. Smaller contributions arise from the two-pion-exchange potential in Fig. 2e (called "TPEP"), which is the analogue of two-photon exchange between charged particles shown in Fig. 2d. There is even an analogue of the atomic polarization force in Fig. 2g, where two electrons simultaneously polarize their nucleus using their electric fields. The nuclear analogue involving three nucleons simultaneously is illustrated in Figs. 2h and 2i, and is called a three-nucleon force [9]. Although relatively weak compared to V_π (a few percent), three-nucleon forces play an important role in fine-tuning nuclear energy levels. The final ingredient is an important short-range interaction (which must be determined by phenomenology) shown in Fig. 2c that has no direct analogue in the physics of light atoms. Just as one can exchange three photons, three-pion-exchange is possible and is depicted in Fig. 2f.

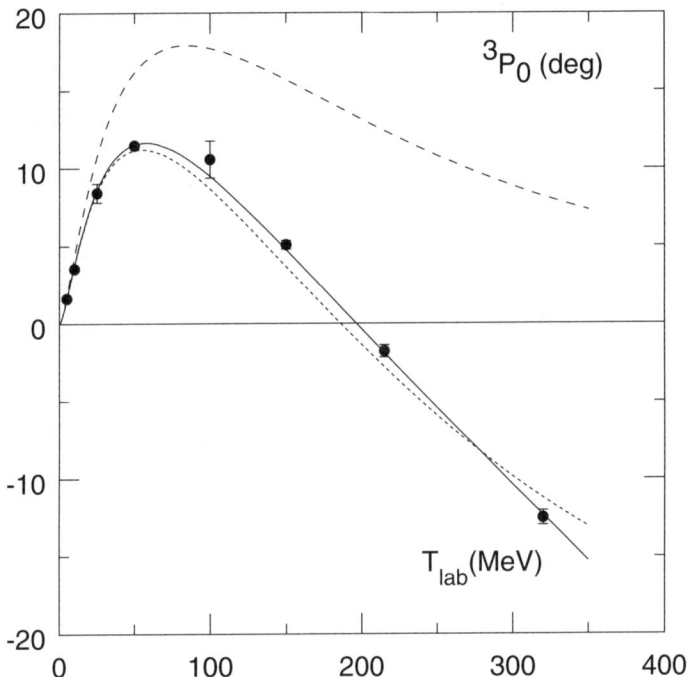

Fig. 3. $^3P_0$ phase shift (in degrees) calculated only with the OPEP tail for $r > b$ (dashed line), and with one (dotted) and three (solid) short-range-interaction terms added. The experimental results are indicated by separate points with error bars [10]

It is worth recalling that the uncertainty principle tells us that exchanging light particles produces longer-range forces and exchanging heavier particles produces shorter-range forces. Thus OPEP has a longer range than TPEP, as illustrated in Fig. 1. Many mechanisms have been proposed for the short-range part of the nuclear force, such as heavy-meson exchange, for example. Any meson-exchange mechanism produces singular forces, which are regularized to make them finite. However one chooses to do this, the part of the nuclear force inside b must be adjusted to fit the nucleon-nucleon scattering data, and no individual parameterization of the short-range force is intrinsically superior (i.e., it doesn't matter how you do it).

How all of this works in practice is indicated in Fig. 3. Imagine that you throw away *all* of the nuclear potential inside $r = b$ (with b chosen to be 1.4 fm) in Fig. 1, keeping only the tail of the force between two nucleons. Now compute a phase shift (the $^3P_0$, for example). This very modest physics input predicts the basic shape of the phase shift (dashed line) as a function of energy. This variation with energy is a consequence of the small pion mass (compared to the energy scale in the figure). What is missing in this curve is a smooth (negative) short-range contribution that grows roughly in proportion to the energy. We can fill in the missing short-range interaction inside $r = b$ by adding a potential term specified by one short-range parameter. This produces the dotted line, which is a rather good fit, and adding two more terms (solid line) produces a nearly perfect fit to the experimental results. Fixing the short-range part of the potential looks very much like making an effective-range expansion. All useful physics is specified by a few parameters, and the details are completely unimportant.

What are the consequences of exchanging a pion rather than a photon? The pseudoscalar nature of the pion mandates its spin-dependent coupling to a nucleon, and this leads to a dominant tensor force between two nucleons. Except for its radial dependence, the form of V_π mimics the interaction between two magnetic dipoles, as seen in the Breit interaction, for example. Thus we have in nuclear physics a situation that is the converse of the atomic case: a dominant tensor force and a smaller central force. In order to grasp the difficulties that nuclear physicists face, imagine that you are an atomic physicist in a universe where magnetic (not electric) forces are dominant, and where QED can be solved only for long-range forces and you must resort to phenomenology to generate the short-range part of the force between electrons and nuclei.

Although this may sound hopeless, it is merely difficult. The key to handling complexities is adequate computing power, and that became routinely available only in the late 1980s or early 1990s. Since then there has been explosive development in our understanding of light nuclei. Underlying all of these developments is an improved understanding of the nuclear force. It is convenient to divide nuclear forces and their history into three distinct time periods.

First-generation nuclear forces were developed prior to 1993. They all contained the one-pion-exchange force, but everything else was relatively crude. The fits to the nucleon-nucleon scattering data (needed to parameterize the short-range part of that force) were indifferent.

Second-generation forces were developed beginning in 1993 [4]. They were more sophisticated and generally very well fit to the scattering data. As an example of how well the fitting worked, the Nijmegen group (which pioneered this sophisticated procedure) allowed the pion mass to vary in the Yukawa function defining V_π, and then fit that mass. They also allowed different masses for the neutral and charged pions that were being exchanged and found [11]

$$m_{\pi^\pm} = 139.4(10)\,\text{MeV}\,, \tag{1}$$

$$m_{\pi^0} = 135.6(13)\,\text{MeV}\,, \tag{2}$$

both results agreeing with free pion masses ($m_{\pi^\pm} = 139.57018(35)$ MeV and $m_{\pi^0} = 134.9766(6)$ MeV [12]). It is both heartening and a bit amazing that the masses of the pions can be determined to better than 1% using data taken in reactions that have no free pions! This result is the best quantitative proof of the importance of pion degrees of freedom in nuclear physics.

Third-generation nuclear forces are currently under development. These forces are quite sophisticated and incorporate two-pion exchange, as well as V_π. All of the pion-exchange forces (including three-nucleon forces) are being generated in accordance with the rules of chiral perturbation theory. One expects even better fits to the scattering data. This is clearly work in progress, but preliminary calculations and versions have already appeared [13].

4   Calculations of Light Nuclei

Having a nuclear force is not very useful unless one can calculate nuclear properties with it. Such calculations are quite difficult. Until the middle 1980s only the two-nucleon problem had been solved with numerical errors smaller than 1%. At that time the three-nucleon systems ^{3}H and ^{3}He were accurately calculated using a variety of first-generation nuclear-force models [14]. Soon thereafter the α-particle (^{4}He) was calculated by Joe Carlson, who pioneered a technique that has revolutionized our understanding of light nuclei: Green's Function Monte Carlo (GFMC) [15].

The difficulty in solving the Schrödinger equation for nuclei is easily understood, although it was not initially obvious. Nuclei are best described in terms of nucleon degrees of freedom. Nucleons come in two types, protons and neutrons, which have nearly the same masses and can be considered as the up and down components of an "isospin" degree of freedom. If one also includes its spin, a single nucleon thus has four internal degrees of freedom. Two nucleons consequently have 16 internal degrees of freedom, which is roughly the number of components in the nucleon-nucleon force (coupling spin, isospin and orbital motion in a very complicated way). To handle this complexity one again requires fast computers, and that is a fairly recent development.

The GFMC technique has been used to solve for all of the bound (and some unbound) states of nuclei with up to 10 nucleons. One member of this collaboration (Steve Pieper [18]) calculated that the ten-nucleon Schrödinger equation requires the solution of more than 200,000 coupled second-order partial-differential equations in 27 continuous variables, and this can be accomplished with numerical errors on the order of 1%! A subset of the results of this impressive calculation are shown in Fig. 4 [19].

Although the nucleon-nucleon scattering data alone can predict the binding energy of the deuteron (^{2}H) to within about 1/2%, the experimental binding energy is used as input data in fitting the nucleon-nucleon potential. The nuclei ^{3}H and ^{3}He (not shown) are slightly underbound without a three-nucleon force, and that force can be adjusted to remedy the underbinding. This highlights both the dominant nature of the nucleon-nucleon force and the relative smallness of three-nucleon forces, which is nevertheless appropriate in size to account for the small discrepancies that result from using only nucleon-nucleon forces in calculations of nuclei with more than two nucleons.

Once the ^{3}H binding energy is fixed, the binding energy of ^{4}He is then accurately predicted to within about 1%. The five-nucleon systems (not shown) are unbound, but their properties are rather well reproduced. The six-nucleon systems are also well predicted. There are small problems with more neutron-rich nuclei (compare ^{9}Li with ^{7}Li or ^{8}He with ^{6}He or ^{4}He), but only 3 adjustable parameters in the three-nucleon force allow several dozen energy levels to be quite well reproduced [19]. Because nuclei are weakly bound systems, there are large cancellations between the (large) potential and (large) kinetic energies, leaving small binding energies. The results shown in Fig. 4 are quite remarkable, espe-

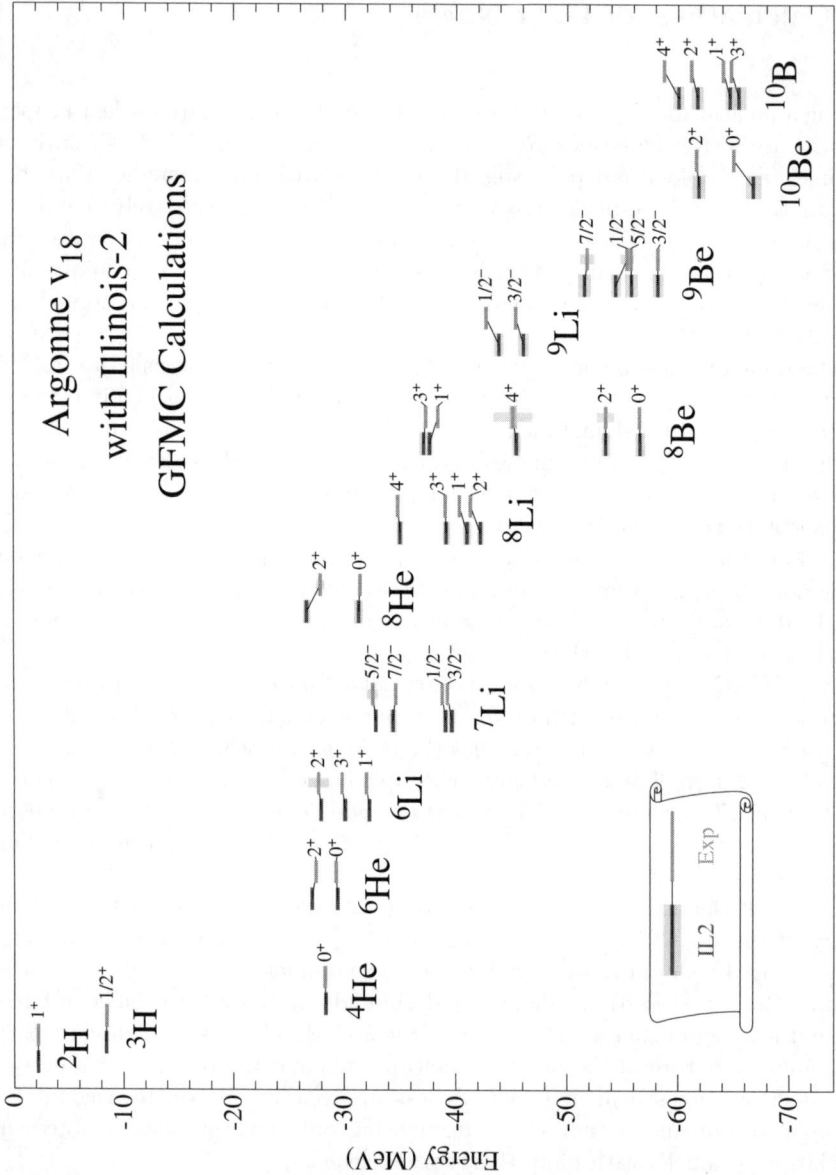

Fig. 4. GFMC calculations of the binding energies of the levels (labelled by their spins and parities) of light nuclei with as many as ten nucleons. These calculations use a common Hamiltonian and have a numerical uncertainty on the order of 1%. Heavy shaded lines to the left are calculated energies (with errors), while light shaded lines to the right are experimental energies. The label "IL2" refers to the Illinois-2 model of the three-nucleon force that is used in all of the calculations together with the Argonne V_{18} two-nucleon force

cially given that small (fractional) discrepancies in the energy components lead to large effects on the binding energies.

We note finally that power counting can be used to show that light nuclei are basically non-relativistic, and relativistic corrections are on the order of a few percent. Power counting is a powerful qualitative technique for determining the relative importance of various mechanisms in nuclear physics.

5   What Nuclear Physics Can Do for Atomic Physics

With our recently implemented computational skills we in nuclear physics can calculate many properties of light nuclei with fairly good accuracy. This is especially true for the deuteron, which is almost unbound and is computationally simple. Although nuclear experiments don't have the intrinsic accuracy of atomic experiments, many nuclear quantities that are relevant to precise atomic experiments can also be measured using nuclear techniques, and usually with fairly good accuracy.

What quantities are we talking about? The nuclear length scale is set by $R \sim 1$ fm $= 10^{-5}$ Å. The much larger atomic length scale of $a_0 \sim 1$ Å means that an expansion in powers of R/a_0 makes great sense, and a typical wavelength for an atomic electron is so large compared to the nuclear size that only moments of the nuclear observables come into play. This also corresponds to an expansion in α, the fine-structure constant, and $m_e R$, where m_e is the electron mass. This is a rapidly converging series.

For processes that have nuclear states inside loops (such as polarizabilities) the excitation energies of those states play a significant role. Although states of any energy can be excited in principle, in practice the effective energy of (virtual) excitation for light nuclei (call it $\bar{\omega}_N$) is within a factor of two of 10 MeV, except for the more tightly bound α-particle, which also has a smaller radius as a consequence. This number follows from the uncertainty principle and the fact that nucleons in a light nucleus have a radius of about 2 fm. The deuteron's weak binding generates the lowest values, which is about 6 MeV for the deuteron's electric polarizability ($\alpha_E \sim 1/\bar{\omega}_N$). Using the value of $\bar{\omega}_N = 10$ MeV, we find $\bar{\omega}_N R \sim 1/10$, which is a reasonably small expansion parameter.

At what levels do various nuclear mechanisms affect the Lamb shift? The (lowest) orders in α that receive nuclear contributions (for S-states) are sketched in Table 1. The various mechanisms are divided into static Coulomb (both non-relativistic and relativistic), recoil (inverse powers of the nuclear mass, M), nuclear structure, vacuum polarization, and radiative processes. The nuclear effects are conveniently divided into two categories: those that directly involve only the properties of the nuclear ground state, and those that involve virtual excited states and are traditionally called "nuclear structure." A radius is a good example of the former, while a polarizability is the prototype of the latter. Calculational techniques are quite different for these two categories.

It is beyond the scope of this review to list detailed formulae and extensive references to past work. I strongly recommend the recent review of [3], which is

Table 1. Orders in α where various contributions to the Lamb shift for S-states have been calculated. The label "f.s." denotes a contribution from nuclear finite size or nuclear structure. Once nuclear physics enters a process at a given order, higher orders will also have nuclear corrections. A "$-$" indicates that although a complete calculation of nuclear contributions has not been made, such contributions are expected. Names refer to the person who first calculated the leading-order term of that type. References and the meanings of other labels are given in the text

Process	α^2	α^4	α^5	α^6
NR Coulomb	Bohr	f.s.	f.s.	f.s.
Rel. Coulomb		Dirac		f.s.
Recoil		Darwin	$-$	$-$
Nucl. Structure			f.s.	$-$
Vacuum Pol.			Uehling	f.s.
Radiative			Bethe	f.s.

extremely well organized. An entire section is devoted to nuclear contributions, and these are listed in their Table 10 with references and numerical values for the hydrogen atom. A sketch of how these contributions scale is given below together with some of the more recent references.

The leading-order non-relativistic energy is simply the Bohr energy of order α^2. Nuclear finite-size contributions of non-relativistic type (i.e., generated by the Schrödinger equation) begin for S-states in order $(Z\alpha)^4$ [20] and are proportional to R^2; they have also been calculated in order $(Z\alpha)^5$ and $(Z\alpha)^6$ [21,22]. The Dirac energy has a leading-order $(Z\alpha)^4$ term, while the nuclear finite-size contributions of relativistic type begin in order $(Z\alpha)^6$ and are proportional to R^2. A recent calculation exists for deuterium [23]. The non-relativistic finite-size corrections of order $(Z\alpha)^5$ and $(Z\alpha)^6$ are tiny for electronic atoms (they contain higher powers of $m_e R$, which is very small), but are not necessarily small for muonic atoms ($m_\mu R$ is about 1 for most light nuclei), which was the original motivation for developing them. P-state finite-size effects begin in order $(Z\alpha)^6$ and are of both relativistic ($\sim R^2$) and non-relativistic ($\sim R^4$) types.

The most important nuclear-structure mechanism is the electric polarizability (which has a long history and will be discussed in more detail later), and this generates a leading contribution of order $\alpha^2(Z\alpha)^3$. Coulomb corrections of order $\alpha^2(Z\alpha)^4$ were developed in the context of a greatly simplified model of the polarizability in muonic atoms [24] (which would not be applicable to electronic atoms).

The Uehling mechanism for vacuum polarization is of order $\alpha(Z\alpha)^4$, while the first nuclear corrections are of order $\alpha(Z\alpha)^5$ [26–29]. The leading-order radiative process is also of order $\alpha(Z\alpha)^4$, while the nuclear finite-size corrections begin in order $\alpha(Z\alpha)^5$ [28,29]. Both of these nuclear corrections are proportional to R^2. Recoil corrections have a long and interesting history that predates the Schrödinger equation (C. G. Darwin derived the leading term of order $(Z\alpha)^4$ using Bohr-Sommerfeld quantization; see the references in [30]). To the best of my knowledge no published calculation exists for the nuclear-finite-size recoil corrections, which begin in order $(Z\alpha)^5$, although the techniques of [31] lead to a result proportional to R^2/M, which should be very small. We note finally the hadronic vacuum polarization, which (although not nuclear in origin) is generated by the strong interactions [32].

One quantity through which nuclear size manifests itself is the nuclear charge form factor (the Fourier transform of the nuclear ground-state charge density, ϱ), which is given by

$$F(\boldsymbol{q}) = \int d^3r\,\varrho(\boldsymbol{r})\,\exp(i\boldsymbol{q}\cdot\boldsymbol{r}) \cong Z(1 - \frac{q^2}{6}\langle r^2\rangle_{\text{ch}} + \cdots) - \tfrac{1}{2}\,q^\alpha q^\beta\,Q^{\alpha\beta} + \cdots, \quad (3)$$

where $\boldsymbol{q}$ is the momentum transferred from an electron to the nucleus, $Q^{\alpha\beta}$ is the nuclear quadrupole-moment tensor, Z is the total nuclear charge, and $\langle r^2\rangle_{\text{ch}}$ is the mean-square radius of the nuclear charge density. These moments should dominate the nuclear corrections to atomic energy levels because $|\boldsymbol{q}|$ in an atom is set by the (very small) atomic scales. Using F to construct the electron-nucleus Coulomb interaction, one obtains

$$V_{\text{C}}(\boldsymbol{r}) \cong -\frac{Z\alpha}{r} + \frac{2\pi Z\alpha}{3}\langle r^2\rangle_{\text{ch}}\,\delta^3(\boldsymbol{r}) - \frac{Q\alpha}{2r^3}\frac{(3\,(\boldsymbol{S}\cdot\hat{\boldsymbol{r}})^2 - \boldsymbol{S}^2)}{S(2S-1)} + \cdots, \quad (4)$$

where $\boldsymbol{S}$ is the nuclear spin operator and Q is the nuclear quadrupole moment (which vanishes unless the nucleus has spin $S \geq 1$). The Fourier transform of the nuclear ground-state current density has a similar expansion

$$\boldsymbol{J}(\boldsymbol{q}) = \int d^3r\,\boldsymbol{J}(\boldsymbol{r})\,\exp(i\boldsymbol{q}\cdot\boldsymbol{r}) \cong -i\boldsymbol{q}\times\boldsymbol{\mu}\,(1 - \frac{q^2}{6}\langle r^2\rangle_{\text{M}} + \cdots) + \cdots, \quad (5)$$

where $\boldsymbol{\mu}$ is the nuclear magnetic moment and $\langle r^2\rangle_{\text{M}}$ is the mean-square radius of the magnetization density. The first term generates the usual atomic hyperfine interaction.

Electron-nucleus scattering experiments are the primary technique used to measure moments of nuclear charge and current densities that are relevant to atomic physics [37], and some appropriate values of these quantities are tabulated in Table 2. An exception is the measurement of the deuteron's quadrupole moment ($Q = 0.282(19)$ fm^2) obtained by scattering polarized deuterons from a high-Z nuclear target at low energy [38]. This result is consistent with the molecular determination ($Q = 0.2860(15)$ fm^2) [39,40], but its error is an order of magnitude larger. Although there is no reason to believe that the (tensor) electric polarizability of the deuteron [45] plays a significant role in the H-D (molecular) quadrupole-hyperfine splitting that was used to determine Q, that correction was not included in the analysis. It was included in the analysis of the nuclear measurement.

I highly recommend the recent review of electron-nucleus scattering by Ingo Sick [37], which contains values of the charge and magnetic radii of light nuclei. That review not only lists the best and most recent values of quantities of interest, but discusses reliability and technical details for those who are interested.

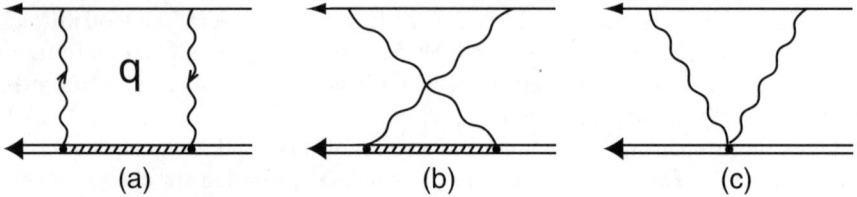

Fig. 5. The direct two-photon process is shown in **a**, the crossed-photon process in **b**, and "seagull" contributions in **c**. The seagulls reflect non-nucleonic processes and terms necessary for gauge invariance. In these graphs the double lines represent a nucleus, the single lines an electron, the wiggly lines a (virtual) photon. The shading represents the set of nuclear excited states. The loop momentum is q, and integrating over this momentum sets the scales of the nuclear part of the process

Table 2. Values of the root-mean-square charge and magnetic radii and the quadrupole moment (if nonvanishing) of the nucleons and various light nuclei obtained by nuclear experiments, together with a selected reference. If two values are given, the second value is that obtained by an atomic or molecular measurement

Nucleus	$\langle r^2 \rangle_{ch}^{1/2}$ (fm)	ref.	$\langle r^2 \rangle_{M}^{1/2}$ (fm)	ref.	Q (fm^2)	ref.
H	0.880 (15)	[33]	0.836 (9)	[34]	–	
	0.883 (14)	[35]			–	
^{2}H	2.130 (10)	[36]	2.072 (18)	[37]	0.282 (19)	[38]
					0.2860 (15)	[39,40]
^{3}H	1.755 (87)	[37]	1.84 (18)	[37]	–	
^{3}He	1.959 (34)	[37]	1.97 (15)	[37]	–	
	1.954 (8)	[41]			–	
^{4}He	1.676 (8)	[42]	–		–	

Nucleon	$\langle r^2 \rangle_{ch}$ (fm^2)	ref.	$\langle r^2 \rangle_{M}^{1/2}$ (fm)	ref.
n	−0.1140 (26)	[43]	0.873 (11)	[44]

One result from that review is listed in Table 2 and is important for the discussion below. The errors of the tritium (^{3}H) radii are nearly an order of magnitude larger than those of deuterium. Of all the light nuclei tritium is the most poorly known experimentally, although the charge radius can now be calculated with reasonable accuracy.

In addition to moments of the nuclear charge and current densities, various components and moments of the nuclear Compton amplitude can play a significant role. Mechanisms that contribute to the polarizabilities are shown in Fig. 5. The direct (sequential) exchange of photons and the crossed-photon process are shown in (a) and (b), while the "seagull" process is shown in (c). The latter mechanism is required by gauge invariance in any model of hadrons with structure. The exchange of pions between nucleons generates such terms, for example [46]. Because these are loop diagrams, they involve an integral over all momenta (q), and this sets the nuclear scales of the problem. The nuclear

Table 3. Values of the electric polarizability of light nuclei, both theoretical and experimental, where the latter have been determined by nuclear experiments. No uncertainties were given for the ^{3}H, ^{3}He, and ^{4}He calculations in [52,53], but they are likely to be smaller than about 10%. The ^{4}He result was used in analyses of muonic He citeHe4-x,He4-t

Nucleus	α_E^{calc} (fm^3)	ref.	α_E^{exp} (fm^3)	ref.
^{2}H	0.6328 (17)	[48]	0.61 (4)	[50]
			0.70 (5)	[51]
^{3}H	0.139	[52]	–	
^{3}He	0.145	[53]	0.250 (40)	[54]
			0.130 (13)	[55]
^{4}He	0.076	[53]	0.072 (4)	[24]

size scale (R), the electron mass (m_e), and the average virtual-excitation energy ($\bar{\omega}_N$, appropriate to the shaded part of the line in (a) and (b) that indicates excited nuclear states) determine the generalized polarizabilities [47]. The process is dominated by the usual electric and magnetic polarizabilities and their logarithmic modifications [48].

Specific examples are the (scalar) electric polarizability, α_E, and the nuclear spin-dependent polarizability ($\sim S$). The latter interacts with the electron spin to produce a contribution to the electron-nucleus hyperfine splitting. There exists a recent calculation of the latter for deuterium [49]. Values of the nuclear electric polarizability for light nuclei obtained from calculations or experiments are listed in Table 3. There are either calculations or measurements of α_E for ^{2}H [48,50,51], ^{3}H [52] and ^{3}He [52–55], and ^{4}He [24,56]. With the exception of the ^{3}He experimental results, there is reasonable consistency. The decreasing size of the polarizabilities for heavier nuclei is caused by their increased binding. The α-particle has more than 10 times the binding energy of the deuteron, and its polarizability is an order of magnitude smaller.

The physics of hyperfine splittings is in general rather different from the physics that contributes to the Lamb shift. It should therefore be no surprise that the nuclear physics that contributes to hyperfine splittings is also quite different; it is also more complicated than its Lamb-shift counterpart. The dominant nuclear physics that we discussed previously was the physics of the nuclear charge density, in the form of moments of the static charge density (i.e., radii) and electric dipole moments that contribute to polarizabilities.

The primary nuclear mechanism in hyperfine splittings is the magnetic interaction caused by the nuclear magnetization density. This density is not as well understood as the charge density. The primary reason is that the same mesons whose exchange binds nuclei together also contribute to the nuclear currents if they carry a charge. The pions that we discussed earlier generate a very important component of that current [46]. The reason for the dichotomy between nuclear charges and currents can be understood by imagining that charged-meson exchange between nucleons is instantaneous. In this limit we know that

Table 4. Difference between hyperfine experiments and QED hyperfine calculations for the nth S-state of light hydrogenic atoms times n^3, expressed as parts per million of the Fermi energy. This difference is interpreted as nuclear contributions to the hyperfine splitting[57]. A negative entry indicates that the theoretical prediction without nuclear corrections is too large

	$n^3(E_{hfs}^{exp} - E_{hfs}^{QED})/E_F$ (ppm)			
State	H	^{2}H	^{3}H	^{3}He$^+$
1S	-33	138	-38	222
2S	-33	137	–	221

the transmitted charge is always on a nucleon. In other words only nucleon degrees of freedom matter, which is the normal situation for the charge density. The power counting that we discussed earlier states that corrections to the nuclear charge operator are small ($\sim 1\%$), and include a type that vanishes for instantaneous meson exchanges. That is not the case for the current, however, since any flow of charge (even from a virtual meson) produces a current that is not simply related to nucleon degrees of freedom, and that current can couple to photons. These meson-exchange currents (often denoted "MEC") can be as large as those generated by the usual nuclear convection current. Various tricks can be used to eliminate part of our ignorance, but the nuclear current density is less well understood than the nuclear charge density. Atomic hyperfine splittings provide us with an excellent opportunity to learn about nuclear currents in a very different setting.

Although most of the hyperfine experiments in light atoms were performed decades ago, there has recently been renewed theoretical interest, and the accuracy of the QED calculations is sufficient to extract nuclear information [57]. The differences between the QED calculations and the experimental results can be interpreted as nuclear corrections, and those are significant, as indicated in Table 4. The S-state results in this table (presented as a ratio) have been taken from Table 1 of the recent work of Ivanov and Karshenboim [57].

Hyperfine structure is generated by short-range interactions. The dominant Fermi contribution (E_F) arises from a δ-function, and that produces a dependence on the square of the electron's nth S-state wave function at the origin, $|\phi_n(0)|^2$, which is proportional to $1/n^3$. Most nuclear effects have the same dependence ($\sim 1/n^3$), which has been removed from the results in Table 4. The 1S and 2S results are seen to be consistent at this level of accuracy, with 1S experimental results typically being much more accurate.

More calculations of the nuclear contributions to hyperfine splittings in light atoms are badly needed if we are to use this information to learn about the currents in light nuclei. These contributions come in the form of Zemach moments [58] (ground-state quantities) and spin-dependent polarizabilities (discussed above). There exists a considerable literature on the latter subject dating back 50 years. The recent work of Mil'shtein and Khriplovich [49] has pointed out a serious defect in that older work. Although the leading-order terms are essentially non-relativistic in origin (for the nucleons in a nucleus), the sub-leading-order terms are not, and require relativity (for the nucleons) in order to obtain a correct result. This is not terribly surprising, since the same physics that enters that polarizability also enters the Gerasimov-Drell-Hearn sum rule [59], which requires relativity at the nucleon level [60], and is a topic of considerable current interest in nuclear physics [61]. The calculations of [49] suggest that deuterium at least can be understood using fairly simple nuclear models. This needs to be checked using more sophisticated models. We note that the Zemach correction [58] adds to the ratio in Table 4, improving the agreement between experiment and theory for H and ^{3}H. The large positive value of that ratio for deuterium suggests a large polarizability correction, which is confirmed by [49].

6   The Proton Size

One recurring problem in the hydrogen Lamb shift is the appropriate value of the mean-square radius of the proton, $\langle r^2 \rangle_p$, to use in calculations. Some older determinations [62] disagree strongly with more recent ones [63]. As shown in (3), the slope of the charge form factor (with respect to q^2) at $q^2 = 0$ determines that quantity. The form factor is measured by scattering electrons from the proton at various energies and scattering angles.

There are (at least) four problems associated with analyzing the charge-form-factor data to obtain the proton size. The first is that the counting rates in such an experiment are proportional to the flux of electrons times the number of protons in the target seen by each electron. That product must be measured. In other words the measured form factor for low q^2 is $(a - b\frac{q^2}{6} + \cdots)$, where $b/a = \langle r^2 \rangle_p$. The measured normalization a (not exactly equal to 1) clearly influences the value and error of $\langle r^2 \rangle_p$. Most analyses unfortunately don't take the normalization fully into account, and [64] estimates that a proper treatment of the normalization of available data could increase $\langle r^2 \rangle_p^{1/2}$ by about 0.015 fm and increase its error, as well. In an atom, of course, the normalization is precisely computable.

Another source of error is neglecting higher-order corrections in α (i.e., Coulomb corrections) and [33] demonstrates that this increases $\langle r^2 \rangle_p^{1/2}$ by about 0.010 fm. A similar problem in analyzing deuterium data was resolved in [36]. Another difficulty that existed in the past was a lack of high-quality low-q^2 data. The final problem is that one must use a sufficiently flexible fitting function to represent $F(q)$, or the errors in the radius will be unrealistically low. All of the older analyses had one or more of these flaws.

Most of the recent analyses [63,33,34] are compatible if the appropriate corrections are made. An analysis by Rosenfelder [33] contains all of the appropriate ingredients, and he obtains $\langle r^2 \rangle_p^{1/2} = 0.880(15)$ fm. There is a PSI experiment now underway to measure the Lamb shift in muonic hydrogen, which would produce the definitive result for $\langle r^2 \rangle_p$ [65,66]. One expects the results of that experiment to be compatible with Rosenfelder's result. Extraction of the proton radius [35] from the electronic Lamb shift is now somewhat uncertain because of controversy involving the two-loop diagrams. These diagrams are significantly less important in muonic hydrogen, where the relative roles of the vacuum polarization and radiative processes are reversed.

7   What Atomic Physics Can Do for Nuclear Physics

The single most valuable gift by atomic physics to the nuclear physics community would be the accurate determination of the proton mean-square radius: $\langle r^2 \rangle_p$. This quantity is important to nuclear theorists who wish to compare their nuclear wave function calculations with measured mean-square radii. In order for an external source of electric field (such as a passing electron) to probe a nucleus, it is first necessary to "grab" the proton's intrinsic charge distribution, which then maps out the mean-square radius of the proton probability distribution in the wave function: $\langle r^2 \rangle_{\mathrm{wfn}}$. Thus the measured mean-square radius of a nucleus, $\langle r^2 \rangle$, has the following components:

$$\langle r^2 \rangle = \langle r^2 \rangle_{\mathrm{wfn}} + \langle r^2 \rangle_p + \frac{N}{Z} \langle r^2 \rangle_n + \frac{1}{Z} \langle r^2 \rangle_{...} , \tag{6}$$

where the intrinsic contribution of the N neutrons has been included as well as that of the Z protons, and $\langle r^2 \rangle_{...}$ is the contribution of everything else, including the very interesting (to nuclear physicists) contributions from strong-interaction mechanisms and relativity in the nuclear charge density [67]. Because the neutron looks very much like a positively charged core surrounded by a negatively charged cloud, its mean-square radius has the opposite sign to that of the proton, whose core is surrounded by a positively charged cloud. It should be clear from (6) that $\langle r^2 \rangle_p$ (which is much larger than $\langle r^2 \rangle_n$) is an important part of the overall mean-square radius. Its present uncertainty degrades our ability to test the wave functions of light nuclei.

The next most important measurements are isotope shifts in light atoms or ions. Since isotope shifts measure differences in frequencies for fixed nuclear charge Z, the effect of the protons' intrinsic size cancels in the difference. This is particularly important given the current lack of a precise value for the proton's radius. The neutrons' effect is relatively small and can be rather easily eliminated, and thus one is directly comparing differences in wave functions, or of small contributions from $\langle r^2 \rangle_{...}$. Isotope shifts are therefore especially "theorist-friendly" measurements, since they are closest to measuring what nuclear theorists actually calculate.

Precise isotope-shift measurements have been performed for ^{4}He - ^{3}He [41] and for ^{2}H - ^{1}H (D-H) [5]. A measurement of ^{6}He - ^{4}He is being undertaken [68] at ANL. Gordon Drake has written about and strongly advocated such measurements in the Li isotopes [69]. These are all highly desirable measurements. Because the ^{3}H (tritium) charge radius currently has large errors, in my opinion the single most valuable measurement to be undertaken for nuclear physics purposes would be the tritium-hydrogen (^{3}H - ^{1}H) isotope shift. An extensive series of calculations using first-generation nuclear forces found $\langle r^2 \rangle_{\mathrm{wfn}}^{1/2}$ for tritium to

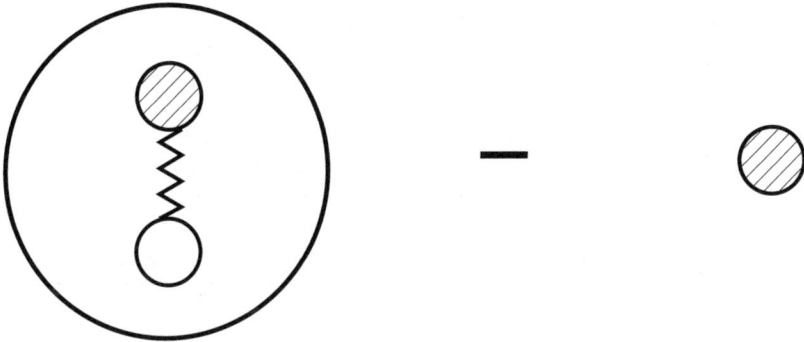

Fig. 6. Cartoon of the ^{2}H - H isotope shift, illustrating how the effect of the finite size of the proton (shaded small circle) in deuterium is cancelled in the measurement. The finite size of the neutron (open small circle) and the electromagnetic interaction mediated by the strong-interaction (binding) mechanism (indicated by the jagged line between the nucleons) do affect the deuteron's charge radius (see text)

be 1.582(8) fm, where the "error" is a subjective estimate [70]. This number could likely be improved by using second-generation nuclear forces, although it will never be as accurate as the corresponding deuteron value, which we discuss next.

The D-H isotope shift in the 2S-1S transition reported by the Garching group [5] was

$$\Delta\nu = 670\ 994\ 334.64(15)\ \text{kHz}. \tag{7}$$

Most of this effect is due to the different masses of the two isotopes (and begins in the first significant figure, indicated by an arrow). The precision is nevertheless sufficiently high that the mean-square-radius effect in the sixth significant figure (second arrow) is much larger than the error. The electric polarizability of the deuteron influences the eighth significant figure, while the deuteron's magnetic susceptibility contributes to the tenth significant figure. It becomes difficult to trust the interpretation of the nuclear physics at about the 1 kHz level, so improving this measurement probably wouldn't lead to an improved understanding of the nuclear physics.

Analyzing this isotope shift and interpreting the residue (after applying all QED corrections) in terms of the deuteron's radius leads to the results [71] in Table 5. The very small binding energy of the deuteron produces a long wave function tail outside the nuclear potential (interpretable as a proton cloud around the nuclear center of mass), which in turn leads to an easy and very accurate calculation of the mean-square radius of the (square of the) wave function. Subtracting this theoretical radius from the experimental deuteron radius (corrected for the neutron's size) determines the effect of $\langle r^2 \rangle_{...}$ on the radius. Although this difference is quite small, it is nevertheless significant and half the size of the error in the corresponding electron-scattering measurement (see Table 2). The high-precision analysis in Table 5 of the content of the deuteron's charge radius

Table 5. Theoretical and experimental deuteron radii for pointlike nucleons. The deuteron wave function radius corresponding to second-generation nuclear potentials and the experimental point-nucleon charge radius of the deuteron (i.e., with the neutron charge radius removed) are shown in the first two columns, followed by the difference of experimental and theoretical results. The difference of the experimental radius with and without the neutron's size is given last for comparison purposes [43]

$\langle r^2 \rangle_{\mathrm{wfn}}^{1/2}$ (fm)	$_{\exp}\langle r^2 \rangle_{\mathrm{pt}}^{1/2}$ (fm)	difference (fm)	$\Delta\langle r^2 \rangle_{\mathrm{n}}^{1/2}$ (fm)
1.9687(18)	1.9753(10)	0.0066(21)	−0.0291(7)

would have been impossible without the precision of the atomic D-H isotope-shift measurement. This measurement has given nuclear physics unique insight into small mechanisms that are at present poorly understood [72].

8   Summary and Conclusions

Nuclear forces and nuclear calculations in light nuclei are under control in a way never before attained. This progress has been possible because of the great increase in computing power in recent years. Many of the nuclear quantities that contribute to atomic measurements have been calculated or measured to a reasonable level of accuracy, a level that is improving with time. Isotope shifts are valuable contributions to nuclear physics knowledge, and are especially useful to theorists who are interested in testing the quality of their wave functions for light nuclei. In special cases such as deuterium these measurements provide the only insight into the size of small contributions to the electromagnetic interaction that are generated by the underlying strong-interaction mechanisms. In my opinion the tritium-hydrogen isotope shift would be the most useful measurement of that type. One especially hopes that the ongoing PSI experiment is successful in measuring the proton size via the Lamb shift in muonic hydrogen.

Acknowledgements

The work of J.L. Friar was performed under the auspices of the United States Department of Energy. The author would like to thank Savely Karshenboim for his interest in nuclear aspects of precise atomic measurements and for giving me the opportunity to discuss this problem.

References

1. D'Arcy Wentworth Thompson: *On Growth and Form* (Cambridge Univ. Press, Cambridge 1959), p. 2
2. J. Reichert, M. Niering, R. Holzwarth, M. Wietz, Th. Udem, and T.W. Hänsch: Phys. Rev. Lett. **84**, 3232 (2000) The H 2S-1S frequency was measured to 14 significant figures in this work

3. M.I. Eides, H. Grotch, and V.A. Shelyuto: Phys. Rep. **342**, 63 (2001) This recent, comprehensive, and very well-organized review is highly recommended
4. V.G.J. Stoks, R.A M. Klomp, C.P.F. Terheggen, and J.J. de Swart: Phys. Rev. C **49**, 2950 (1994); R.B. Wiringa, V.G.J. Stoks, and R. Schiavilla: Phys. Rev. C **51**, 38 (1995); J.L. Friar, G.L. Payne, V.G.J. Stoks, and J.J. de Swart: Phys. Lett. B **311**, 4 (1993)
5. A. Huber, Th. Udem, B. Gross, J. Reichert, M. Kourogi, K. Pachucki, M. Weitz, and T.W. Hänsch: Phys. Rev. Lett. **80**, 468 (1998)
6. S. Weinberg: Physica A **96**, 327 (1979); in *A Festschrift for I.I. Rabi*, Transactions of the N.Y. Academy of Sciences **38**, 185 (1977); Nucl. Phys. B **363**, 3 (1991); Phys. Lett. B **251**, 288 (1990); Phys. Lett. B **295**, 114 (1992)
7. U. van Kolck: Thesis, University of Texas, (1993); C. Ordóñez and U. van Kolck: Phys. Lett. B **291**, 459 (1992); C. Ordóñez, L. Ray, and U. van Kolck: Phys. Rev. Lett. **72**, 1982 (1994); U. van Kolck: Phys. Rev. C **49**, 2932 (1994); C. Ordóñez, L. Ray, and U. van Kolck: Phys. Rev. C **53**, 2086 (1996)
8. J.L. Friar: Few-Body Systems **22**, 161 (1997)
9. S.A. Coon, M.D. Scadron, P.C. McNamee, B.R. Barrett, D.W.E. Blatt, and B.H.J. McKellar: Nucl. Phys. A **317**, 242 (1979)
10. V.G.J. Stoks, R.A.M. Klomp, M.C.M. Rentmeester, and J.J. de Swart: Phys. Rev. C **48**, 792 (1993)
11. R.A.M. Klomp, V.G.J. Stoks, and J.J. de Swart: Phys. Rev. C **44**, R1258 (1991); V. Stoks, R. Timmermans, and J.J. de Swart: Phys. Rev. C **47**, 512 (1993)
12. K. Hagiwara, et al.: Phys. Rev. D **66**, 010001 (2002)
13. M.C. M. Rentmeester, R.G.E. Timmermans, J.L. Friar, J.J. de Swart: Phys. Rev. Lett. **82**, 4992 (1999); M. Walzl, U.-G. Meißner, and E. Epelbaum: Nucl. Phys. A **693**, 663 (2001); D.R. Entem and R. Machleidt: Phys. Lett. B **524**, 93 (2002)
14. C.R. Chen, G.L. Payne, J.L. Friar, and B.F. Gibson: Phys. Rev. C **31**, 2266 (1985); J.L. Friar, B.F. Gibson, and G.L. Payne: Phys. Rev. C **35**, 1502 (1987)
15. J. Carlson and R. Schiavilla: Rev. Mod. Phys. **70**, 743 (1998) This is a comprehensive review of light nuclei containing an elementary discussion of GFMC. What Joe Carlson pioneered in the late 1980s was the application of GFMC to solving the nuclear problem with realistic potentials, where difficulties of principle exist (the so-called "fermion problem"). The GFMC technique was pioneered for nuclear physics with very simplified potentials in [16]. Atomic and molecular applications began about 1967 [17]
16. G.A. Baker, Jr., J.L. Gammel, B.J. Hill, and J.G. Wills: Phys. Rev. **125**, 1754 (1962); M. H. Kalos: Phys. Rev. **128**, 1791 (1962)
17. M. Kalos: J. Comp. Phys. **1**, 257 (1967)
18. S.C. Pieper: Private Communication
19. S.C. Pieper, K. Varga, and R. B. Wiringa: Phys. Rev. C **66** (2002) [in press] This is the latest in a series of calculations of light nuclei. Details and references for the nuclear forces that they used can be found here
20. R. Karplus, A. Klein, and J. Schwinger: Phys. Rev. **86**, 288 (1952)
21. J.L. Friar: Phys. Lett. B **80**, 157 (1979); J. L. Friar: Ann. Phys. (N.Y.) **122**, 151 (1979)
22. E.E. Trofimenko: Phys. Lett. A **73**, 383 (1979); L.A. Borisoglebsky and E.E. Trofimenko: Phys. Lett. B **81**, 175 (1979)
23. J.L. Friar and G.L. Payne: Phys. Rev. A **56**, 5173 (1997)
24. J.L. Friar: Phys. Rev. C **16**, 1540 (1977). This work used the results of [25]
25. Y.M. Arkatov, et al.: Yad. Fiz. **19**, 1172 (1974) [Sov. J. Nucl. Phys. **19**, 598 (1974)]

26. J.L. Friar: Z. Phys. A **292**, 1 (1979); (E) A **303**, 84 (1981)
27. D.J. Hylton: Phys. Rev. A **32**, 1303 (1985)
28. M.I. Eides and H. Grotch: Phys. Rev. A **56**, R2507 (1997)
29. K. Pachucki: Phys. Rev. A **48**, 120 (1993). See the discussion in [3]
30. J.L. Friar and J.W. Negele: Phys. Lett. B **46**, 5 (1973)
31. H. Grotch and D.R. Yennie: Rev. Mod. Phys. **41**, 350 (1969) This work provides the basis for treating reduced-mass and recoil corrections
32. J.L. Friar, J. Martorell, and D.W.L. Sprung: Phys. Rev. A **59**, 4061 (1999) This is only the latest work on this interesting topic. It contains results of previous calculations together with references
33. R. Rosenfelder: Phys. Lett. B **479**, 381 (2000)
34. P. Mergell, U.-G. Meißner, and D. Drechsel: Nucl. Phys. A **596**, 367 (1996)
35. K. Melnikov and T. van Ritbergen: Phys. Rev. Lett. **84** 1673 (2000) They find $\langle r^2 \rangle_{\mathrm{p}}^{1/2} = 0.883(14)$ fm
36. I. Sick and D. Trautmann: Phys. Lett. B **375**, 16 (1996)
37. I. Sick: Prog. Part. Nucl. Phys. **47**, 245 (2001) This is an excellent review and is highly recommended
38. J.E. Kammeraad and L. D. Knutson: Nucl. Phys. A **435**, 502 (1985)
39. R.V. Reid, Jr. and M.L. Vaida: Phys. Rev. Lett. **29**, 494 (1972); (E) **34**, 1064 (1975)
40. D.M. Bishop and L.M. Cheung: Phys. Rev. A **20**, 381 (1979)
41. D. Shiner, R. Dixson, and V. Vedantham: Phys. Rev. Lett. **74**, 3553 (1995) The value deduced in their Ref. [22] was used
42. I. Sick: Phys. Lett. B **116**, 212 (1982)
43. S. Kopecky, P. Riehs, J.A. Harvey, and N.W. Hill: Phys. Rev. Lett. **74**, 2427 (1995). The compiled value at the bottom of Table 1 was used
44. G. Kubon, et al.: Phys. Lett. B **524**, 26 (2002)
45. M.H. Lopes, J. A. Tostevan, and R. C. Johnson: Phys. Rev. C **28**, 1779 (1983)
46. J.L. Friar: Phys. Rev. Lett. **36**, 510 (1976)
47. J.L. Friar and G.L. Payne: Phys. Rev. C **56** 619 (1997)
48. J.L. Friar and G.L. Payne: Phys. Rev. C **55**, 2764 (1997) See the references in this work for earlier calculations of α_{E}
49. A.I. Mil'shtein and I.B. Khriplovich: JETP **82**, 616 (1996) [Zh. Eksp. Teor. Fiz. **109**, 1146 (1996)]
50. J.L. Friar, S. Fallieros, E.L. Tomusiak, D. Skopik, and E.G. Fuller: Phys. Rev. C **27**, 1364 (1983)
51. N.L. Rodning, L.D. Knutson, W.G. Lynch, and M.B. Tsang: Phys. Rev. Lett. **49**, 909 (1982)
52. V.D. Efros, W. Leidemann, and G. Orlandini: Phys. Lett. B **408**, 1 (1997). Two of their calculations of α_{E} for ^{3}H were averaged in our Table 3
53. W. Leidemann: in *Proceedings of the XVIIIth European Conference on Few-Body Problems in Physics*, Few-Body Syst. (suppl.) (in press); Private Communication
54. F. Goeckner, L.O. Lamm, and L.D. Knutson: Phys. Rev. C **43**, 66 (1991)
55. G.A. Rinker: Phys. Rev. A **14**, 18 (1976) He estimated a 10% error in his α_{E} calculation for ^{3}He
56. R. Rosenfelder: Nucl. Phys. A **393**, 301 (1983)
57. S.G. Karshenboim and V. G. Ivanov: Phys. Lett. B **524**, 259 (2002)
58. C. Zemach: Phys. Rev. **104**, 1771 (1956)
59. S.D. Drell and A.C. Hearn: Phys. Rev. Lett. **16**, 908 (1966); S.B. Gerasimov: Sov. J. Nucl. Phys. **2**, 430 (1966)

60. J.L. Friar: Phys. Rev. C **16**, 1540 (1977)
61. H.R. Weller: in *Proceedings of GDH 2000*, ed. by D. Drechsel and L. Tiator (World Scientific, Singapore 2001), p. 145
62. L.N. Hand, D.J. Miller, and R. Wilson: Rev. Mod. Phys. **35**, 335 (1963) The value of $\langle r^2 \rangle_{\mathrm{p}}^{1/2} = 0.805(11)$ fm is not reliable and should not be used
63. G.G. Simon, Ch. Schmitt, F. Borkowski, V.H. Walter: Nucl. Phys. A **333**, 381 (1980)
64. C.W. Wong: Int. J. Mod. Phys. E **3**, 821 (1994)
65. D. Taqqu, et al.: Hyperfine Int. **119**, 311 (1999)
66. S.G. Karshenboim: Can. J. Phys. **77**, 241 (1999)
67. J.L. Friar: Czech. J. Phys. **43**, 259 (1993); H. Arenhövel: Czech. J. Phys. **43**, 207 (1993) These are introductory articles treating meson-exchange currents (strong-interaction contributions to electromagnetic currents). Neglecting such currents and other binding effects is commonly denoted the "impulse approximation"
68. Z.-T. Lu: Private Communication
69. Z.-C. Yan and G.W.F. Drake: Phys. Rev. A **61**, 022504 (2000)
70. J.L. Friar: in *XIVth International Conference on Few-Body Problems in Physics*, ed. by F. Gross, AIP Conference Proceedings **334**, 323 (1995) Theoretical values of $\langle r^2 \rangle_{\mathrm{wfn}}^{1/2}$ for ^{3}He and ^{3}H were obtained from the fits in Fig. 1. The ^{3}He value of 1.769(5) fm was used in the talk
71. J.L. Friar, J. Martorell, and D.W.L. Sprung: Phys. Rev. A **56**, 4579 (1997). A tiny center-of-mass contribution to the deuteron radius was overlooked in that work, but has been included in Table 5 above. I would like to thank Ingo Sick for pointing out the problem
72. A.J. Buchmann, H. Henning, and P.U. Sauer: Few-Body Systems **21**, 149 (1996)

Deeply Bound Pionic States as an Indicator of Chiral Symmetry Restoration

Toshimitsu Yamazaki

RI Beam Science Laboratory, RIKEN, Wako, Saitama, 351-0198 Japan

Abstract. A review is made on the recently developed pion-nucleus bound-state spectroscopy and its implication. Whereas pionic-atom x-ray spectroscopy cannot access to deeply bound π^- states in heavy nuclei (such as the 1s states in Pb), a new type of nuclear spectroscopy, using so called pion transfer reactions, has been developed at GSI using the $(d,{}^3\text{He})$ reaction at 500–600 MeV deuteron energy. Recent experiments on the 1s π^- states in ${}^{205}\text{Pb}$ and Sn isotopes have shown that the isovector Parameter, bi, is enhanced over the free-π^-N value, indicating that the chiral Order Parameter (f_π^2) is reduced in the nuclear medium by a significant amount.

1   Pionic Atoms – Old History and New Frontier

Pionic atoms and their spectroscopy have been known for many years [1]. Negative pions are captured by nuclei into pionic atom states at around $\approx \sqrt{m_\pi/m_e}$ (≈ 17), and cascade down to lower excited states. The pionic x-ray spectroscopy of cascade transitions yielded many fruitful data on the strong-interaction shifts and widths, which occur near the end of the cascades. However, the traditional pionic-atom spectroscopy has never touched the inner Orbits inside so called last orbitals, where the atomic cascade terminates simply because the nuclear absorption competes with the radiative process:

$$\Gamma_{\text{absorption}} > \Gamma_{\text{radiation}} \quad [\sim \text{kev region}]. \tag{1}$$

This does not mean at all that there is no discrete state inside the last orbital, as the above cascade termination condition is nothing to do with the *quasi-stability* condition of a level:

$$\Gamma_{absorption} < \Delta E_{n,n-1} \quad [\sim \text{MeV region}]. \tag{2}$$

The quasi-stability of π^- in heavy nuclei was pointed out by Friedman and Soff [2] as being due to the repulsive π^- interaction. The calculation of the π^- energy levels in ${}^{90}\text{Zr}$ and ${}^{208}\text{Pb}$ by Toki et al. [3,4] showed discrete energy levels down to the 1s states, as shown in Fig. 1. Surprisingly, no attention had been paid to the possibility of nuclear spectroscopy of this inner world before Toki et al. [3,4] clarified that the deeply bound π^- states in heavy nuclei tan be formed by *pion-transfer reactions*. The π^- mesons in deeply bound states are *not brought from outside* as in ordinary pionic atoms, but they are *produced from inside*. Thus, the deeply bound states possess dual characters; as nuclear resonance states at around 135 MeV as well as π^- bound states below the pion

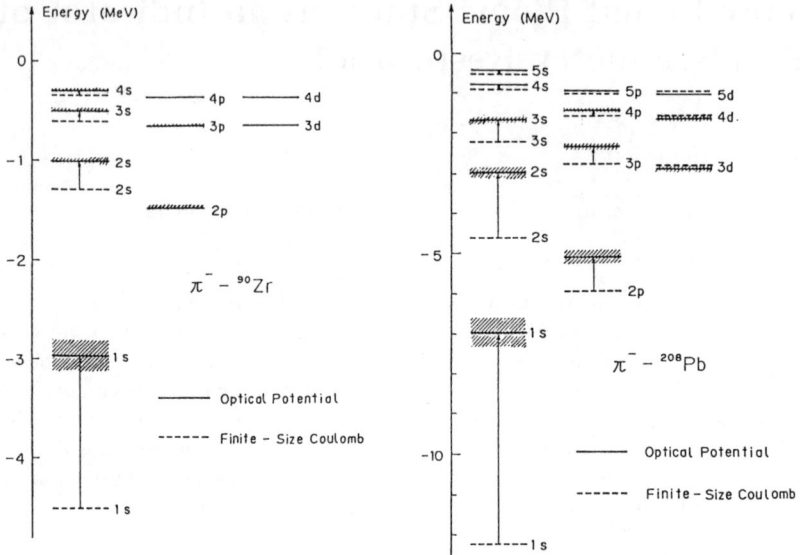

Fig. 1. The calculated energy levels and widths of π^- bound by ^{90}Zr and ^{208}Pb [3]

mass (140 MeV), The present review describes the evolution of this new type of nuclear spectroscopy and its connection with the important issue of chiral symmetry restoration in nuclear medium [5,6].

2 Prediction for Quasi-Stable Pionic Nuclei

The narrowness of deeply bound pionic states, even the 1s states in ^{208}Pb, is understood in terms of a Coulomb pocket which is accommodated by the lon-grange attractive Coulomb interaction and the short-range repulsive π-nucleus interaction, as shown in Fig. 2. The π^- is pushed outward from the center of the nucleus by the strong repulsive interaction, thus forming a halo like state. These states provide unique experimental information as to how the s-wave part of the pion-nucleon interaction is modified in the nuclear medium, whereas the traditional pionic-atom x-ray spectroscopy [l] gives information on shallow bound states only, which are related to both the s-wave and p-wave parts.

3 Observation of Deeply Bound Pionic States

Production of deeply bound pionic states by the (d, ^{3}He) reaction was proposed and studied theoretically by Toki et al. [7,8]. The first successful (d, ^{3}He) experiment was carried out at GSI, using a 600-MeV deuteron beam on a ^{208}Pb target [9–121. The predicted 1s and 2p states were identified as discrete states, which are embedded in the nuclear continuum at around 138 MeV, as shown in Fig. 3. Since this reaction converts one of the neutrons occupying the shallow

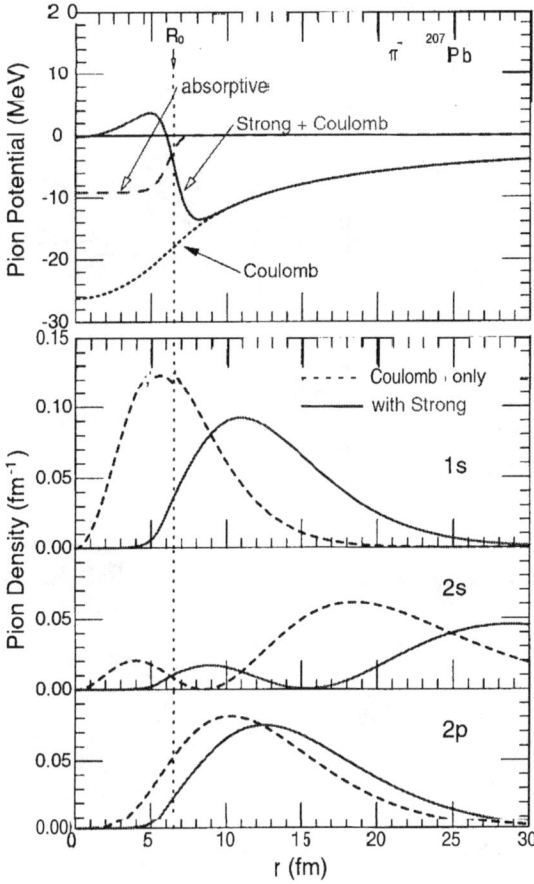

Fig. 2. The π^--^{208}Pb potential and the density distributions of the deeply bound 1s, 2s and 2p π^- states [3]

$3p_{1/2}$ and $3p_{3/2}$ orbitals, the best populated pionic state was $(2p)_\pi$, $(3p_{1/2,3/2})_n^{-1}$ at this kinematical condition where the momentum transfer is very small. The (1s), state was produced as a small satellite.

The second experiment was focused on a ^{206}Pb target to observe a clearly separated 1s peak [13], as demonstrated in Fig. 4. The binding energy and width of the 1s π^- state in ^{205}Pb were used to deduce the potential parameters.

A recent experiment [16] showed preferentially populated distinct $(ls)_\pi$, peaks in three Sn isotopes (Fig. 5), in which the $3s_{1/2}$ orbital is the shallowest lying neutron orbital. This experiment was carried out with much attention on the absolute calibration of the energy scale. For this purpose the normal protonpickup reactions, $^A\mathrm{Sn}(d, 3\mathrm{He})^{A-1}\mathrm{In}(\mathrm{g.s.})$, was employed.

Furthermore, a thin mylar film was deliberately deposited on the downstream surface of each Sn target so that a monoenergetic peak arising from

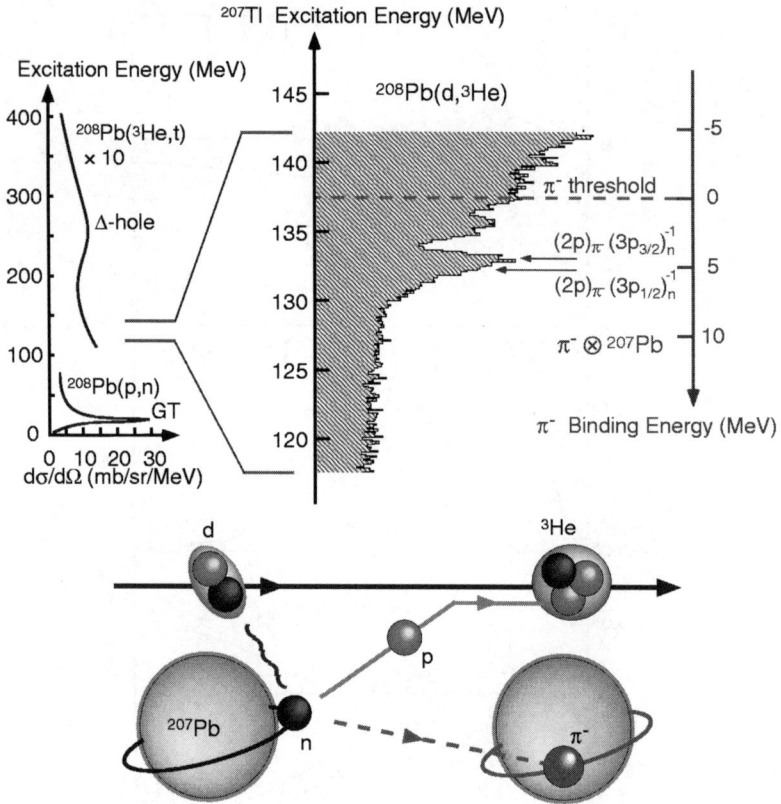

Fig. 3. (*Upper*) The observed ^{208}Pb (d, ^{3}He) reaction spectrum revealing the predicted deeply bound π^- states. (*Lower*) Mechanism to produce π^- states from "inside"

the p(d, ^{3}He)π^0 reaction served as a built-in calibration. In this way, the uncertainty in the binding energy was minimized to $\sim$15 keV. The obtained values of the binding energies and widths of the 1s states in the three Sn isotopes are shown in Fig. 6.

4   Pion-Nucleus Interaction

The π^- binding energies of the 1s π^- states depend on the local (s-wave) part of the poin-nucleus potential,

$$U_s(r) = \frac{2\pi}{m_\pi}[\epsilon_1\{b_0\rho(r) + b_1\Delta\rho(r)\} + \epsilon_2 B_0\rho(r)^2], \qquad (3)$$

where $\rho(r) = \rho^{(n)}(r) + \rho^{(p)}$ and $\Delta\rho(r) = \rho^{(n)}(r) - \rho^{(p)}(r)$, and $\epsilon_1 = 1 + M_\pi/M = 1.15$ and $\epsilon_2 = 1 + m_\phi/2M = 1.07$. It is well known that the 1s states of π^-/ bound in both light and heavy nuclei are governed mainly by the repulsive s-wave Part,

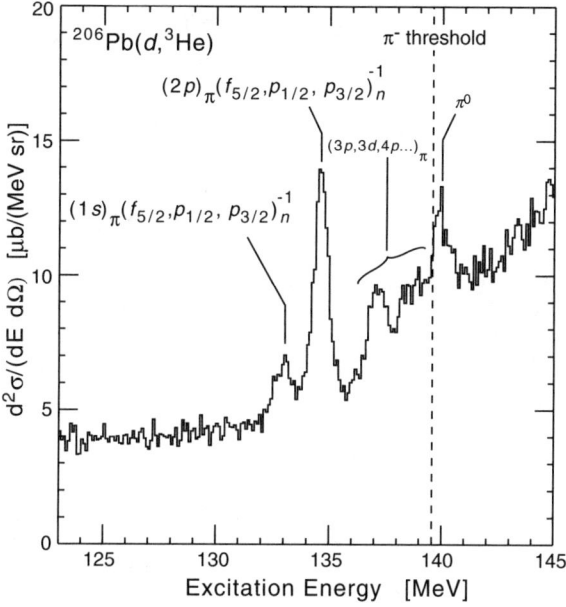

Fig. 4. Double differential Cross section measured in the ^{206}Pb(d, ^{3}He) reaction at incident deuteron energy $T_d = 604.3$ MeV as a function of the excitation energy above the ^{205}Pb ground state. The peak at 140 MeV designated as "π^0" is due to the reaction p(d, ^{3}He)π^0

whereas higher-lying states ($l > 0$) are affected by both the repulsive s-wave and the attractive p-wave parts, which produce more or less cancelling effects. The newly found 1s states in heavy $N > Z$ nuclei are of particular importante in giving the isoscalar interaction Parameter (b_0) and the isovector Parameter (b_1) separately.

The binding energy and width of a π^- state (n, l) are used to deduce the potential Parameters. First, let us examine which part (r) of the nuclear density, $\rho(r)$, is probed by a bound π^-. Two typical nuclei, ^{16}O and ^{208}Pb, are considered. The former possesses only shallow bound states, whereas the latter accommodates deeply bound states. The nuclear matter densities, $\rho(r)$, are assumed to take a 2-parameter Fermi distribution with a half-density radius and a diffuseness parameter. The proton densities are taken from a compilation of Fricke et al. [19] and the neutron distributions are from a systematics obtained by Trzcińska et al. [20]. We take known potential Parameters for the p-wave parts [1] and a set of s-wave Parameters ($b_0 = -0.028$, $b_1 = -0.12$, Re$B_0 = 0$ and Im$B_0 = 0.055$).

Figure 7 Shows the π^- densities ($R_{nl}(r)^2$), the nuclear densities ($\rho(r)$) and the overlapping densities (namely, the nuclear densities probed by π^-), defined as

$$S(r) = \rho(r) \left| R_{nl}(r) \right|^2 r^2, \tag{4}$$

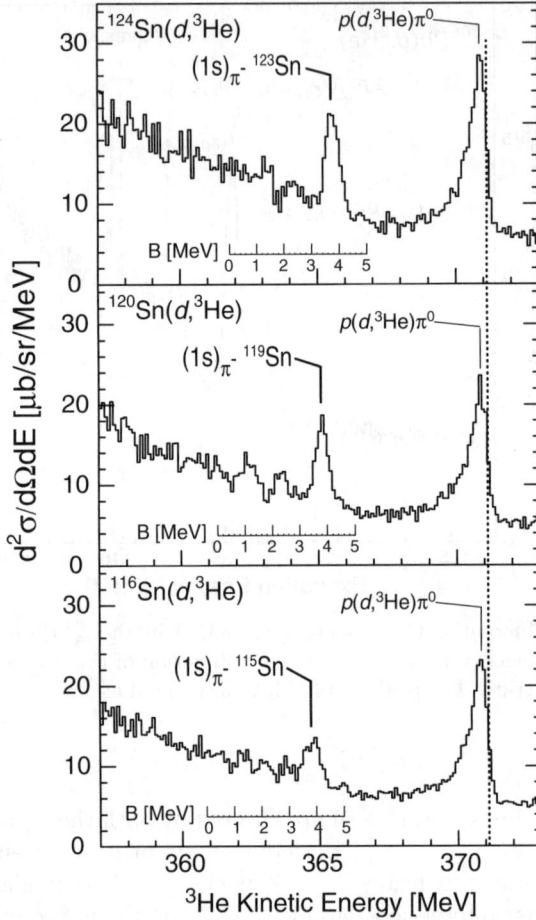

Fig. 5. ^{124,120j116}Sn$(d,^3\mathrm{He})$ reaction spectra measured at $T_d = 503.388$ MeV. The scales of the π^- binding energies in ^{123,119,115}Sn are also indicated. The skewed peak at the right-hand side from the p$(d, 3\mathrm{He})\pi^0$ reaction is used for an absolute calibration of the energy scale

for ^{16}O (1s and 2p) and ^{208}Pb (ls, 2p and 3d). From these figures we notice that the overlapping density is peaked at a radius slightly less than the half-density radius, nearly independent of the nucleus and the π^- quantum numbers. This means that the bound π^- effectively probes a fraction of the full nuclear density $(\rho_0 = \rho(0))$

$$\rho_e \sim 0.60\,\rho_0. \tag{5}$$

There are many potential Parameters in the Ericson-Ericson potential which consists of the s-wave and p-wave Parts [17,18]. So called global fits to all the available empirical data have been made to determine the parameters [I]. However, these parameters are mutually correlated, and the individual Parameters obtained in global fits are subject to large uncertainties, although the parame-

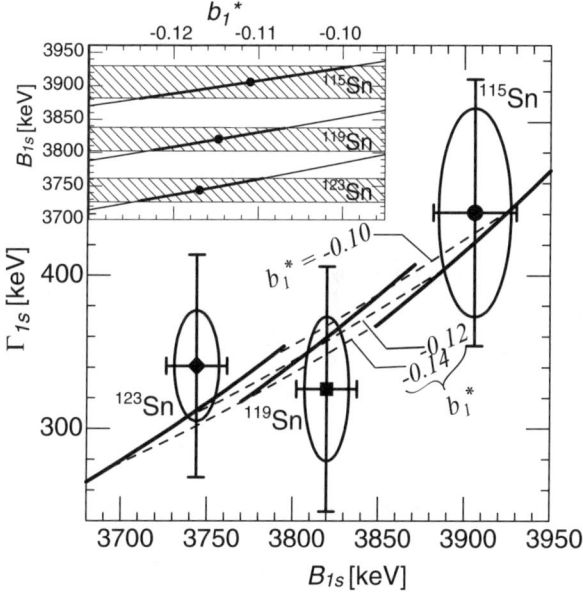

Fig. 6. Experimental values of B_{is}, and Γ_{1s} for π^- in ^{115,119,123}Sn, shown together with theoretical relations for the three isotopes with b_1^* as a parameter and with $\mathrm{Im}B_0$ fixed to $0.046\ m_\pi^{-14}$. The corresponding values of b_1^* on each of the curves tan be read from the dashed lines. (Inset) The theoretical relations between B_{1s}, and b_1^* for the three ^{115,119,123}Sn isotopes are shown with experimental zones

ters tan reproduce the data very well. To determine one particular parameter with an important physical meaning (namely, b_1, in the present case) we need another type of approach developed in [14]. First, we select only the 1s π^- states to determine the s-wave potential part. They are available either in light pionic atoms up to ^{28}Si or as deeply bound states in heavy nuclei. To determine the isoscalar part we select only the symmetric nuclei ($N = Z$). Because of a strong correlation between the $b_0\rho(r)$ and the $\mathrm{Re}B_0\rho(r)^2$ terms we plot best-fit values of b_0 versus a given $\mathrm{Re}B_0$ value in Fig. 8, which gives a correlated relation represented by

$$b_0^* = b_0 + 0.215\mathrm{Re}B_0 = -0.0280 \pm 0.0010. \qquad (6)$$

Such a relation, first emphasized by Seki and Masutani [21], is a result of the overlapping pion density, as shown and extended recently in [22]. For comparison the listed results from representative global fits are shown in the same figure. They are very much distributed. It is clear that the isoscalar strength is not represented by individual b_0 and $\mathrm{Re}B_0$, but by the combined parameter b_0^0 as given above.

Then, we obtain an empirical constraint between b_0 and b_1 using the experimental binding energy and width of the 1s π^- state in ^{205}Pb, as given in Fig. 8. A detailed explanation is given in [14]. The free-πN values ("FREE") [27,28]

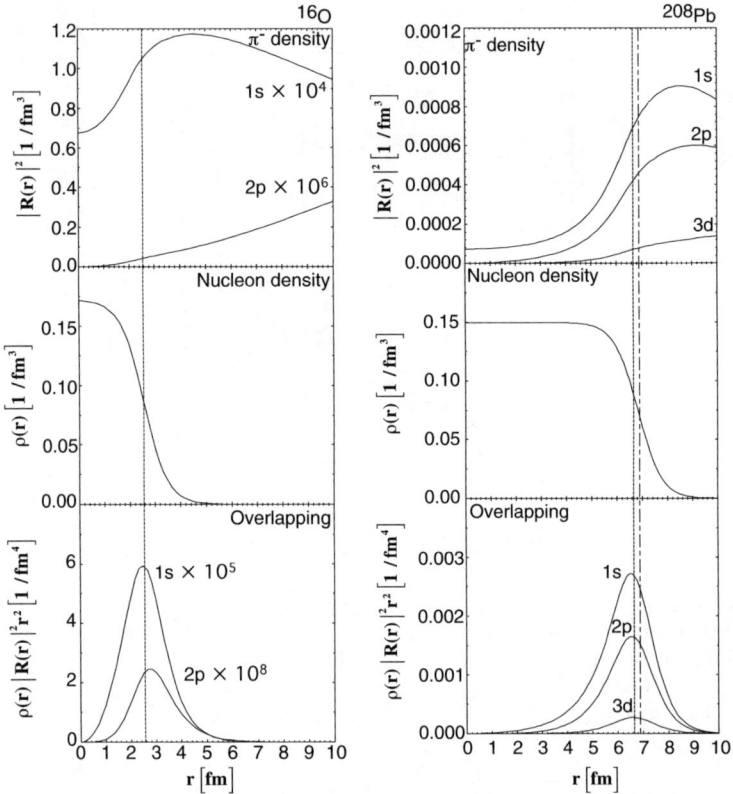

Fig. 7. Overlapping probabilities (*lower frame*) of the π^- densities (*upper frame*) with the nucleon densities (*middle frame*) in typical pionic bound states: (*Left*) ^{16}O. (*Right*) ^{208}Pb. The vertical broken lines show the half-density proton radii and the vertical dash-dotted line is for the half-density neutron radius in ^{208}Pb

and their effective values ("EFF") after the double scattering corrections [17] are also indicated. The cross zone are found to be far from the effective Zone, indicating

$$b_1 = -0.116^{+0-015}_{-0.017}\, m_p^{-1}. \tag{7}$$

The b_0^* value is also enhanced, largely arising from the enhanced b_1. Its comparison with the curve from the double-stattering effect yields

$$\mathrm{Re}B_0 = \frac{b_0^* - b_0^{\mathrm{DS}}}{0.215} \sim -0.038 \pm 0.025, \tag{8}$$

which is small compared with the large values as so far invoked "missing repulsion" [1].

The new data of the Sn experiment, as shown in Fig. 6, yielded values of b_1^*. They are presented in Fig. 10.

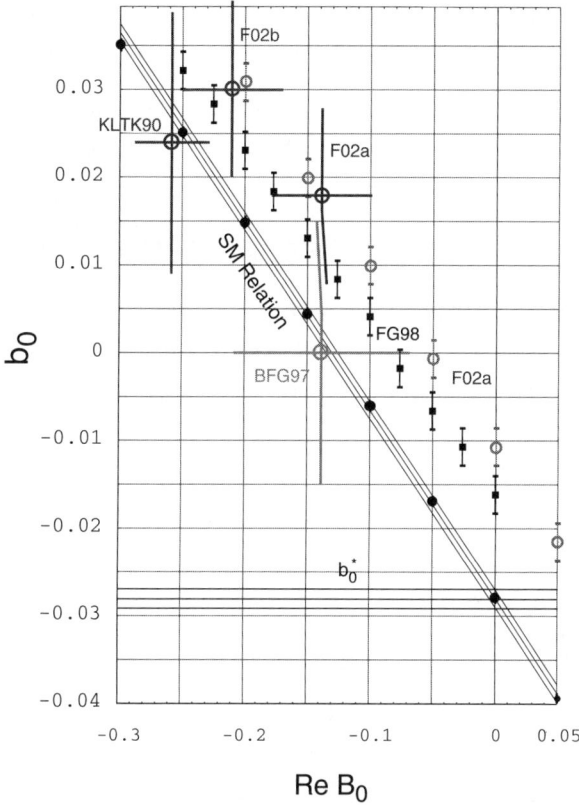

Fig. 8. Seki-Masutani relation between b_0 and $\mathrm{Re}B_0$. Best-fit values of b_0 versus $\mathrm{Re}B_0$ as a gridding variable, obtained in χ^2 minimization using the 1s pionic atom data in 6 $N = 2$ nuclei, are shown by closed circles, whose sizes are equal to the fitting uncertainties. They lie on the SM lines: $b_0^* = b_0 + 0.215\mathrm{Re}B_0 = -0.0280 \pm 0.0010$. The $\{b_0, \mathrm{Re}B_0\}$ parameters obtained from global fits of KLTK9O [23], BFG97 [1], F02a [25] and one of F02b [26] are shown by large open circles with both vertical and horizontal error bars. Also shown are $\{b_0, \mathrm{Re}B_0\}$ sets from FG98 [24] (*closed squares*) and F02a [25] (*small open circles*), without conversion from b_0 to b_0^F in their convention

5   Evidence for In-Medium Restoration of Chiral Symmetry

Chiral symmetry breaking is known to account for the large hadron masses despite the very small u- and d-quark masses [29,30]. Partial restoration of chiral symmetry is expected in nuclear media [30–32], but so far, there has been no clear evidente for this exciting scenario. So called invariant mass spectroscopy for investigating "in-medium" hadron masses of hadrons in unbound nuclear media has an inherent difficulty due to the collisional broadening and shift, as pointed out by us [33]. In this connection a new light has been shed on the pion-nucleus interaction, since the isovector s-wave part is related to the long standing issue of possible chiral symmetry restoration in nuclear medium [5,6]. It is obvious that the information from deeply bound states in heavy nuclei plays an essential role.

The chiral low-energy theorem by Tomozawa [34] and Weinberg [35] connects the pion-nucleon T-matrices directly with the pion decay constant f_π ($= 92.4$ MeV),

$$\text{isoscalar}: \quad TC^{(+)} = \frac{1}{2}[T_{\pi^- p} + T_{\pi^- n}] = 4\pi(l + m_\pi/M)b_0 = 0 \tag{9}$$

$$\text{isovector}: \quad TC^{(+)} = \frac{1}{2}[T_{\pi^- p} + T_{\pi^- n}] = -4\pi(l + m_\pi/M)b_1 = \frac{m_\pi}{2f_\pi^2} \tag{10}$$

The f_π is the order parameter of chiral symmetry breaking of QCD and is expressed in terms of a quark condensate $< \bar{q}q >$ through the Gell-Mann-Oakes-Renner relation [36]

$$m_\pi^2 f_\pi^2 = -m_q < \bar{q}q >, \tag{11}$$

where $m_q = (m_u + m_d)/2$ is the average of the u and d quark masses. Reduction of the chiral Order Parameter (the pion decay constant, f_π) in the nuclear medium is theoretically expected [37–391].

The observed enhancement of b_1^* over the free π^- value ($-0.090\, m_\pi^{-1}$, recently obtained from the experiment on the 1s π^- states in ^{205}Pb and Sn isotopes,

$$b_1^* = -0.115 \pm 0.005\, m_\pi^{-1}, \tag{12}$$

is related to a reduction of the chiral order parameter [37,39]:

$$\frac{b_1^{\text{free}}}{b_1^*} = \frac{f_\pi^*(\rho_e)^2}{f_\pi^2} = 0.78 \pm 0.03. \tag{13}$$

Since the π^- probes the nuclear density of $\rho_{\text{eff}} \sim 0.60\rho_0$ [22], the above value is interpreted as

$$\frac{f_\pi^*(\rho_0)^2}{f_\pi^2} \approx \frac{< \bar{q}q >_{\rho_0}}{< \bar{q}q >_0} = 0.65 \pm 0.05 \tag{14}$$

at the normal nuclear density, $\rho = \rho_0$. This value is very close to predicted values (~ 0.65) [5,40,38]. In this way we have obtained an *evidence for partial restoration of chiral symmetry from weil-defined bound states in a weil defined nuclear density*.

Acknowledgements

The author would like to thank the members of the GSI S160-236 experiments for the fruitful collaboration and H. Toki, S. Hirenzaki, R.S. Hayano, P. Kienle, W. Weise and T. Ericson for their stimulating discussions. Supports by Grantsin-Aid of Japanese Monbukagakusho and Japan Society for the Promotion of Science are gratefully acknowledged.

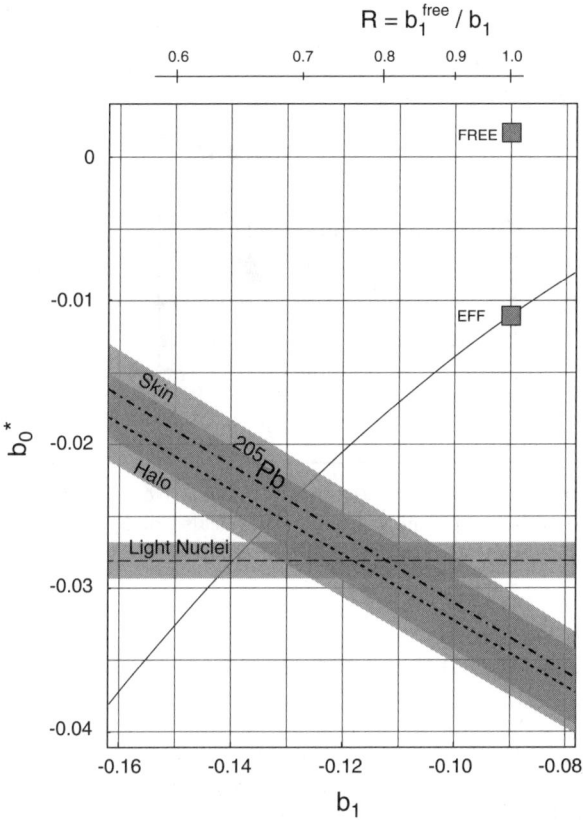

Fig. 9. Presentation of the isoscalar (b_0^*) and the isovector (b_1) parameters. The 1s states of light symmetric pionic atoms give a shaded horizontal band. The $b_0^* - b_1$ constraints by the 1s state of ^{205}Pb are shown for the two types of neutron distributions: ??halo type" (*broken line*) and "skin type" (*dash-dotted line*) with respective 1σ error zones. The free-π^-N values ("FREE") and their effective values ("EFF") after the double scattering corrections are also indicated. The scale for the reduction of the squared chiral order parameter $R = b_1^{\text{free}}/b_1 \approx f_\pi^{*2}/f_\pi^2$ is shown in the upper part

References

1. See C. Batty, E. Riedman and A. Gal: Phys. Reports **287**, 385 (1997) and references therein
2. E. Friedman and G. Soff: J. Phys. G, Nucl. Phys. **11**, L37 (1985)
3. H. Toki and T. Yamazaki: Phys. Lett. B **213**, 129 (1988)
4. H. Toki, S. Hirenzaki, R.S. Hayano and T. Yamazaki: Nucl. Phys. A **501**, 653 (1989)
5. W. Weise: Acta Physica Polonica 31, (2000) 2715; Nucl. Phys. A **690**, 98 (2001)
6. P. Kienle and T. Yamazaki: Phys. Lett. B **514**, 1 (2001)
7. H. Toki, S. Hirenzaki and T. Yamazaki: Nucl. Phys. A **530**, 679 (1991)
8. S. Hirenzaki, H. Toki and T. Yamazaki: Phys. Rev. C **44**, 2472 (1991)

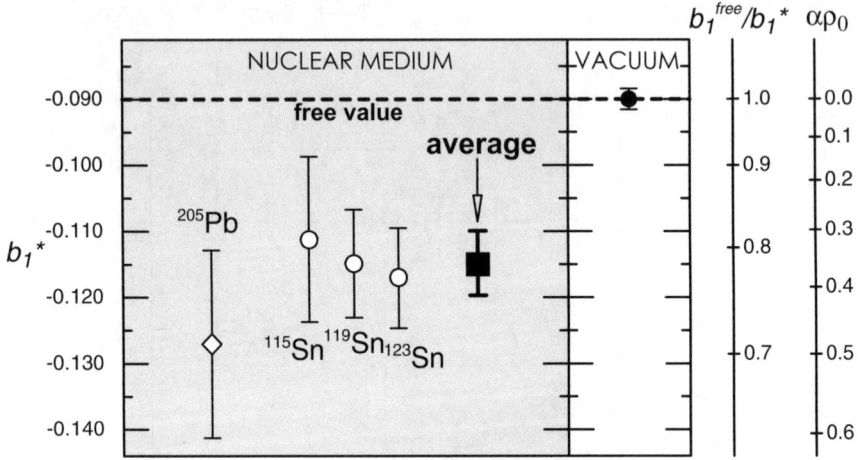

Fig. 10. Summary of the b_1^* values (in m_π^{-1}) deduced from the B_{1s}, binding energies in ^{115,119,123}sN together with the corresponding scales in b_1^{free}/b_1^* ($= f1 *_\pi (\rho_e)^2/f_\pi^2$) and $\alpha\rho_9$. The previous ^{205}Pb data reanalyzed with $\text{Im}B_0(= 0.046\,m_\pi^{-4})$ is shown for comparison

9. T. Yamazaki et UZ.: Z. Phys. A **355**, 219 (1996)
10. T. Yamazaki et UZ.: Phys. Lett. B **418**, 246 (1998)
11. H. Gilg et UZ.: Phys. Rev. C **62**, 025201 (2000)
12. K. Itahashi et UZ.: Phys. Rev. C **62**, 025202 (2000)
13. H. Geissel et UZ.: Phys. Rev. Lett. **88**, 122301 (2002)
14. H. Geissel et UZ.: Phys. Lett. B **549**, 64 (2002)
15. Y. Umemoto, S. Hirenzaki, K. Kume and H. Toki: Phys. Rev. C **62**, 024606 (2000)
16. K. Suzuki et UZ.: nucl-ex/0211023
17. M. Ericson and T.E.O. Ericson: Ann. Phys. **36**, 323 (1966)
18. T.E.O. Ericson and W. Weise: "Pions in Nuclei" (Clarendon Press, Oxford, 1988)
19. G. Fricke, C. Bernhardt, K. Heilig, L.A. Schaller, L. Schellenberg, E.B. Shera and C.W. de Jager: At. Nucl. Data Tables **60**, 177 (1995)
20. A. Trzcinska, J. Jastrzebski, P. Lubinski, F. J. Hartmann, R. Schmidt, T. von Egidy, B. Klos: Phys. Rev. Lett. **87**, 082501 (2001)
21. R. Seki and K. Masutani: Phys. Rev. C **27**, 2799 (1983)
22. T. Yamazaki and S. Hirenzaki: nucl-th/0210040vl
23. J. Konijn, C.T.A.M. de Laat, A. Taal and J.H. Koch: Nucl. Phys. A **519**, 773 (1990)
24. E. Friedman and A. Gal: Phys. Lett. B **432**, 235 (1998)
25. E. Friedman: Phys. Lett. B **524**, 87 (2002)
26. E. Friedman: Nucl. Phys. A **710**, 117 (2002)
27. H.-Ch. Schröder et UZ.: Phys. Lett. B **469**, 25 (1999)
28. T.E.O. Ericson, B. Loiseau and A.W. Thomas: Nucl. Phys. A **684**, 380 (2001)
29. Y. Nambu and G. Jona-Lasinio: Phys. Rev. **122**, 345 (1961)
30. See T. Hatsuda and T. Kunihiro: Phys. Reports **247**, 221 (1994) and references therein
31. U. Vogl and W. Weise: Prog. Part. Nucl. Phys. **27**, 195 (1991)

32. G.E. Brown and M. Rho: Phys. Rep. **269**, 333 (1996); G.E. Brown, C.H. Lee, M. Rho and V. Thorsson: Nucl. Phys. A **567**, 937 (1994)
33. T. Yamazaki and Y. Akaishi: Phys. Lett. B **453**, 1 (1999)
34. Y. Tomozawa: Nuovo Ciment0 A **46**, 707 (1966)
35. S. Weinberg: Phys. Rev. Lett. **17**, 616 (1966)
36. M. Gell-Mann, R.J. Oakes and B. Renner: Phys. Rev. **175**, 2195 (1968)
37. Y. Thorsson and A. Wirzba: Nucl. Phys. A **589**, 633 (1995)
38. See U.-G. Meissner, J.A. Oller and A. Wirzba: Ann. Phys. **297**, 27 (2002) and references therein
39. E.E. Kolomeitsev, N. Kaiser and W. Weise: nucl-th/0207090
40. E.G. Drukarev and E.M. Levin: Nucl. Phys. A **511**, 679 (1990)

Part III

Hydrogen-Like Ions

Virial Relations for the Dirac Equation and Their Applications to Calculations of Hydrogen-Like Atoms

Vladimir M. Shabaev

Department of Physics, St. Petersburg State University
Oulianovskaya Street 1, Petrodvorets, St. Petersburg 198504, Russia

Abstract. Virial relations for the Dirac equation in a central field and their applications to calculations of H-like atoms are considered. It is demonstrated that using these relations allows one to evaluate various average values for a hydrogenlike atom. The corresponding relations for non-diagonal matrix elements provide an effective method for analytical evaluations of infinite sums that occur in calculations based on using the reduced Coulomb-Green function. In particular, this method can be used for calculations of higher-order corrections to the hyperfine splitting and to the g factor in hydrogenlike atoms.

1 Introduction

In non-relativistic quantum mechanics, the virial theorem for a particle moving in a central field $V(r)$ is given by the well known equation:

$$\langle T \rangle = \frac{1}{2} \langle r \frac{dV}{dr} \rangle , \tag{1}$$

where $\langle T \rangle$ denotes the average value of the kinetic energy in a stationary state. This theorem can easily be derived from the equation [1]

$$\frac{d}{dt} \langle A \rangle = \frac{i}{\hbar} \langle [H, A] \rangle = 0 , \tag{2}$$

if we take $A = (\mathbf{r} \cdot \mathbf{p})$. Equation (2), which is generally called as the hypervirial theorem, can also be employed to derive virial relations for diagonal matrix elements of other operators [2]. An extension of these relations to the case of non-diagonal matrix elements was considered in [3,4].

For the Dirac equation in a central field, the virial theorem was first derived by Fock [5]. If we denote the bound-state energy by E, it gives

$$E = \langle mc^2 \beta \rangle + \langle r \frac{dV}{dr} \rangle + \langle V \rangle , \tag{3}$$

where β is the Dirac matrix. For the Coulomb field, one easily finds

$$E = \langle mc^2 \beta \rangle . \tag{4}$$

Some additional virial relations for the Dirac equation were obtained by a number of authors (see, e.g., [2,6] and references therein). Virial relations, which yield recurrence formulas for various average values, were obtained in [2,7–9]. In the case of the Coulomb field, these relations can be employed to derive explicit formulas for the average values $\langle r^s \rangle$, $\langle r^s \beta \rangle$, and $\langle ir^s(\boldsymbol{\alpha} \cdot \mathbf{n})\beta \rangle$, where $\boldsymbol{\alpha}$ is a vector incorporating the Dirac matrices, $\mathbf{n} = \mathbf{r}/r$, and s is integer. The corresponding recurrence relations for non-diagonal matrix elements were derived in [9]. In the case of the Coulomb field, it was found that these relations can be employed to derive explicit formulas for the first-order corrections to the Dirac wave function due to interaction with perturbative potentials of the form $\sim r^s$, $r^s \beta$, $r^s(\boldsymbol{\alpha} \cdot \mathbf{n})$, and $ir^s(\boldsymbol{\alpha} \cdot \mathbf{n})\beta$. Later on (see references below), this method was used for calculations of various corrections to the energy levels, to the hyperfine structure splitting, and to the bound-electron g factor. In contrast to direct analytical and numerical calculations, the virial relation method allows one to derive formulas for various physical quantities by simple algebraic transformations.

In the present paper, following mainly to [9], we derive the virial relations for the Dirac equation and examine their applications to calculations of H-like atoms. Relativistic units ($\hbar = c = 1$) are used in the paper.

2   Derivation of the Virial Relations for the Dirac Equation

For the case of a central field $V(r)$, the Dirac equation has the form

$$(-i\boldsymbol{\alpha} \cdot \boldsymbol{\nabla} + \beta m + V(r))\psi(\mathbf{r}) = E\psi(\mathbf{r}) \,. \tag{5}$$

The wave function is conveniently represented by

$$\psi(\mathbf{r}) = \begin{pmatrix} g(r)\Omega_{\kappa m}(\mathbf{n}) \\ if(r)\Omega_{-\kappa m}(\mathbf{n}) \end{pmatrix} \,, \tag{6}$$

where $\kappa = (-1)^{j+l+1/2}(j + 1/2)$ is the quantum number determined by the angular momentum and the parity of the state. Substituting this expression into (5), we obtain the radial Dirac equations

$$\frac{dG}{dr} + \frac{\kappa}{r}G - (E + m - V)F = 0 \,, \tag{7}$$

$$\frac{dF}{dr} - \frac{\kappa}{r}F + (E - m - V)G = 0 \,, \tag{8}$$

where $G(r) = rg(r)$ and $F(r) = rf(r)$. Introducing the operator [10]

$$H_\kappa = -i\sigma_y \frac{d}{dr} + \sigma_x \frac{\kappa}{r} + \sigma_z m + V \,, \tag{9}$$

where σ_x, σ_y, and σ_z are the Pauli matrices, and denoting

$$\phi(r) = \begin{pmatrix} G(r) \\ F(r) \end{pmatrix} \,, \tag{10}$$

we obtain

$$H_\kappa \phi = E\phi . \tag{11}$$

The operator H_κ is self-adjoint in the space of two-component functions satisfying the boundary conditions

$$\phi(0) = \phi(\infty) = 0 . \tag{12}$$

The scalar product in this space is defined by

$$\langle a|b \rangle = \int_0^\infty dr \, (G_a G_b + F_a F_b) . \tag{13}$$

Let us denote the eigenvalues and the eigenvectors of the operator H_κ by $E_{n\kappa}$ and $\phi_{n\kappa}$, respectively, where n is the principal quantum number. Taking into account the self-adjointness of H_κ, we can write down the following equations

$$\langle n'\kappa'|(H_{\kappa'}Q - QH_\kappa)|n\kappa \rangle = (E_{n'\kappa'} - E_{n\kappa})\langle n'\kappa'|Q|n\kappa \rangle , \tag{14}$$

$$\langle n'\kappa'|(H_{\kappa'}Q + QH_\kappa)|n\kappa \rangle = (E_{n'\kappa'} + E_{n\kappa})\langle n'\kappa'|Q|n\kappa \rangle . \tag{15}$$

Substituting $Q = r^s$, $i\sigma_y r^s$ into equation (14) and $Q = \sigma_z r^s$, $\sigma_x r^s$ into equation (15), and using the commutation properties of the Pauli matrices, we obtain [9]

$$(E_{n'\kappa'} - E_{n\kappa})\langle n'\kappa'|r^s|n\kappa \rangle = -s\langle n'\kappa'|i\sigma_y r^{s-1}|n\kappa \rangle$$
$$+ (\kappa' - \kappa)\langle n'\kappa'|\sigma_x r^{s-1}|n\kappa \rangle , \tag{16}$$

$$(E_{n'\kappa'} - E_{n\kappa})\langle n'\kappa'|i\sigma_y r^s|n\kappa \rangle = s\langle n'\kappa'|r^{s-1}|n\kappa \rangle$$
$$- (\kappa' + \kappa)\langle n'\kappa'|\sigma_z r^{s-1}|n\kappa \rangle + 2m\langle n'\kappa'|\sigma_x r^s|n\kappa \rangle , \tag{17}$$

$$(E_{n'\kappa'} + E_{n\kappa})\langle n'\kappa'|\sigma_z r^s|n\kappa \rangle = s\langle n'\kappa'|\sigma_x r^{s-1}|n\kappa \rangle$$
$$- (\kappa' - \kappa)\langle n'\kappa'|i\sigma_y r^{s-1}|n\kappa \rangle + 2m\langle n'\kappa'|r^s|n\kappa \rangle$$
$$+ 2\langle n'\kappa'|\sigma_z V r^s|n\kappa \rangle , \tag{18}$$

$$(E_{n'\kappa'} + E_{n\kappa})\langle n'\kappa'|\sigma_x r^s|n\kappa \rangle = -s\langle n'\kappa'|\sigma_z r^{s-1}|n\kappa \rangle$$
$$+ (\kappa' + \kappa)\langle n'\kappa'|r^{s-1}|n\kappa \rangle + 2\langle n'\kappa'|\sigma_x V r^s|n\kappa \rangle . \tag{19}$$

In the next sections, we apply these equations for calculations of the average values of various physical quantities as well as for calculations of various higher-order corrections.

3   Application of the Virial Relations for Evaluation of the Average Values

Let consider equations (16)-(19) for the Coulomb field ($V(r) = -\alpha Z/r$) and for diagonal matrix elements ($n'\kappa' = n\kappa$). Denoting

$$A^s = \int_0^\infty dr \, r^s (G_{n\kappa}^2 + F_{n\kappa}^2) , \tag{20}$$

$$B^s = \int_0^\infty dr\, r^s (G_{n\kappa}^2 - F_{n\kappa}^2)\,, \tag{21}$$

$$C^s = 2\int_0^\infty dr\, r^s G_{n\kappa} F_{n\kappa}\,, \tag{22}$$

we obtain [2,7,9]

$$2mA^s - 2E_{n\kappa}B^s = 2\alpha Z B^{s-1} - sC^{s-1}\,, \tag{23}$$
$$2mC^s = -sA^{s-1} + 2\kappa B^{s-1}\,, \tag{24}$$
$$2E_{n\kappa}C^s = 2\kappa A^{s-1} - sB^{s-1} - 2\alpha Z C^{s-1}\,. \tag{25}$$

From these equations, one easily finds

$$-[(s+1)E_{n\kappa} + 2\kappa m]A^s + [(s+1)m + 2\kappa E_{n\kappa}]B^s + 2\alpha Z m C^s = 0 \tag{26}$$

Using equations (23)-(26) for $s = 0, 1$, we obtain

$$B^0 = \frac{E_{n\kappa}}{m}\,, \qquad C^0 = \frac{\kappa}{\alpha Z}\frac{m^2 - E_{n\kappa}^2}{m^2}\,, \qquad B^{-1} = \frac{m}{\alpha Z}\frac{m^2 - E_{n\kappa}^2}{m^2}\,. \tag{27}$$

In addition, according to the Hellmann-Feynman theorem, we have

$$\frac{\partial E_{n\kappa}}{\partial \kappa} = \langle n\kappa | \frac{\partial H_\kappa}{\partial \kappa} | n\kappa \rangle = \langle n\kappa | \frac{\sigma_x}{r} | n\kappa \rangle\,. \tag{28}$$

It yields

$$C^{-1} = \frac{\partial E_{n\kappa}}{\partial \kappa} = \frac{(\alpha Z)^2 \kappa m}{N^3 \gamma}\,, \tag{29}$$

where $N = \sqrt{(\gamma + n_r)^2 + (\alpha Z)^2}$, $\gamma = \sqrt{\kappa^2 - (\alpha Z)^2}$, and $n_r = n - |\kappa|$. The derivative with respect to κ in equation (29) must be taken at a fixed n_r. Using the formulas for B^0, C^0, B^{-1}, and C^{-1} given above and reccurence equations (23)-(26), we can calculate the integrals A^s, B^s, and C^s for any integer s. Explicit formulas for these calculations were derived in [7]. The formulas expressing the integrals A^{s+1}, B^{s+1}, and C^{s+1} in terms of the integrals A^s, B^s, and C^s are

$$\begin{aligned} A^{s+1} = &\{2\alpha Z E_{n\kappa}(s+1)A^s + 2\alpha Z m(s+2)B^s \\ &- (s+1)[sm + 2(m + \kappa E_{n\kappa})]C^s\}\{2(s+2)(m^2 - E_{n\kappa}^2)\}^{-1}\,, \end{aligned} \tag{30}$$
$$\begin{aligned} B^{s+1} = &\{2\alpha Z m(s+1)A^s + 2\alpha Z E_{n\kappa}(s+2)B^s \\ &- (s+1)[2\kappa m + (s+2)E_{n\kappa}]C^s\}\{2(s+2)(m^2 - E_{n\kappa}^2)\}^{-1}\,, \end{aligned} \tag{31}$$
$$C^{s+1} = \frac{1}{2m}[2\kappa B^s - (s+1)A^s]\,. \tag{32}$$

The reversed formulas are

$$A^s = \frac{4\alpha Z(s+2)(mB^{s+1} - E_{n\kappa}A^{s+1})}{(s+1)[(s+1)^2 - 4\gamma^2]} + \frac{(s+1)m + 2\kappa E_{n\kappa}}{\alpha Z m[(s+1)^2 - 4\gamma^2]}$$

$$\times\{[(s+2)m + 2\kappa E_{n\kappa}]B^{s+1} - [(s+2)E_{n\kappa} + 2\kappa m]A^{s+1}\}, \tag{33}$$

$$B^s = \frac{4\alpha Z(mA^{s+1} - E_{n\kappa}B^{s+1})}{(s+1)^2 - 4\gamma^2} - \frac{(s+1)E_{n\kappa} + 2\kappa m}{\alpha Z m[(s+1)^2 - 4\gamma^2]}$$

$$\times\{[(s+2)E_{n\kappa} + 2\kappa m]A^{s+1} - [(s+2)m + 2\kappa E_{n\kappa}]B^{s+1}\}, \tag{34}$$

$$C^s = \frac{1}{2\alpha Z m}[(s+1)E_{n\kappa} + 2\kappa m]A^s - \frac{1}{2\alpha Z m}[(s+1)m + 2\kappa E_{n\kappa}]B^s. \tag{35}$$

Employing these formulas, one easily finds

$$C^1 = \frac{2\kappa E_{n\kappa} - m}{2m^2}, \tag{36}$$

$$A^{-1} = \frac{\alpha Z m}{\gamma N^3}(\kappa^2 + n_r\gamma), \tag{37}$$

$$B^{-1} = \frac{m^2 - E_{n\kappa}^2}{\alpha Z m}, \tag{38}$$

$$A^{-2} = \frac{2(\alpha Z)^2\kappa[2\kappa(\gamma + n_r) - N]m^2}{N^4(4\gamma^2 - 1)\gamma}, \tag{39}$$

$$B^{-2} = \frac{2(\alpha Z)^2[2\gamma^2 N - \kappa(\gamma + n_r)]m^2}{N^4(4\gamma^2 - 1)\gamma}, \tag{40}$$

$$C^{-2} = \frac{2(\alpha Z)^3[2\kappa(\gamma + n_r) - N]m^2}{N^4(4\gamma^2 - 1)\gamma}, \tag{41}$$

$$A^{-3} = \frac{2(\alpha Z)^3 m^3}{N^5(4\gamma^2 - 1)\gamma(\gamma^2 - 1)}$$
$$\times[N^2(1 + 2\gamma^2) - 3\kappa N(\gamma + n_r) + 3(\alpha Z)^2(N^2 - \kappa^2)]. \tag{42}$$

It should be noted here that the integral A^{-3} exists only for $|\kappa| \geq 2$. The integral C^1 occurs in calculations of the bound-electron g factor. The integrals C^{-2} and A^{-3} occur in calculations of the magnetic dipole and electric quadrupole hyperfine splitting, respectively, (see [11] for details). Formulas (27)-(29) and (36)-(41) were also employed in calculations of the recoil corrections to the atomic energy levels [12,13].

4   Application of the Virial Relations for Calculations of Higher-Order Corrections

In calculations of higher-order corrections to various physical quantities one needs to evaluate the sums

$$|i, s, \kappa', n\kappa\rangle \equiv \sum_{n'}^{(E_{n'\kappa'} \neq E_{n\kappa})} \frac{|n'\kappa'\rangle\langle n'\kappa'|R_i^s|n\kappa\rangle}{E_{n\kappa} - E_{n'\kappa'}}, \tag{43}$$

where $R_1^s = r^s$, $R_2^s = \sigma_z r^s$, $R_3^s = \sigma_x r^s$, and $R_4^s = i\sigma_y r^s$. For instance, to derive the first-order correction to the hydrogenic wave function due to the magnetic dipole hyperfine interaction, we need to evaluate the expression (43) for R_3^{-2}. Let us consider how virial relations (16)-(19) can be employed for calculations of these sums in the case of a hydrogenlike atom ($V(r) = -\alpha Z/r$). For this case, equations (16)-(19) can be rewritten in the following form

$$(E_{n\kappa} - E_{n'\kappa'})A_{n'\kappa',n\kappa}^s = sD_{n'\kappa',n\kappa}^{s-1} + (\kappa - \kappa')C_{n'\kappa',n\kappa}^{s-1}, \tag{44}$$

$$(E_{n\kappa} - E_{n'\kappa'})D_{n'\kappa',n\kappa}^s = -2mC_{n'\kappa',n\kappa}^s - sA_{n'\kappa',n\kappa}^{s-1} + (\kappa' + \kappa)B_{n'\kappa',n\kappa}^{s-1}, \tag{45}$$

$$(E_{n\kappa} - E_{n'\kappa'})B_{n'\kappa',n\kappa}^s = -2mA_{n'\kappa',n\kappa}^s + 2E_{n\kappa}B_{n'\kappa',n\kappa}^s - sC_{n'\kappa',n\kappa}^{s-1}$$
$$+ (\kappa' - \kappa)D_{n'\kappa',n\kappa}^{s-1} + 2\alpha Z B_{n'\kappa',n\kappa}^{s-1}, \tag{46}$$

$$(E_{n\kappa} - E_{n'\kappa'})C_{n'\kappa',n\kappa}^s = 2E_{n\kappa}C_{n'\kappa',n\kappa}^s - (\kappa' + \kappa)A_{n'\kappa',n\kappa}^{s-1} + sB_{n'\kappa',n\kappa}^{s-1}$$
$$+ 2\alpha Z C_{n'\kappa',n\kappa}^{s-1}, \tag{47}$$

where

$$A_{n'\kappa',n\kappa}^s = \int_0^\infty dr \, r^s (G_{n'\kappa'}G_{n\kappa} + F_{n'\kappa'}F_{n\kappa}), \tag{48}$$

$$B_{n'\kappa',n\kappa}^s = \int_0^\infty dr \, r^s (G_{n'\kappa'}G_{n\kappa} - F_{n'\kappa'}F_{n\kappa}), \tag{49}$$

$$C_{n'\kappa',n\kappa}^s = \int_0^\infty dr \, r^s (G_{n'\kappa'}F_{n\kappa} + F_{n'\kappa'}G_{n\kappa}), \tag{50}$$

$$D_{n'\kappa',n\kappa}^s = \int_0^\infty dr \, r^s (G_{n'\kappa'}F_{n\kappa} - F_{n'\kappa'}G_{n\kappa}). \tag{51}$$

From equations (44)-(47), we obtain

$$(E_{n\kappa} - E_{n'\kappa'})(E_{n\kappa}D_{n'\kappa',n\kappa}^s + mC_{n'\kappa',n\kappa}^s)$$
$$= [-sE_{n\kappa} - m(\kappa' + \kappa)]A_{n'\kappa',n\kappa}^{s-1} + [sm + E_{n\kappa}(\kappa' + \kappa)]B_{n'\kappa',n\kappa}^{s-1}$$
$$+ 2\alpha Z m C_{n'\kappa',n\kappa}^{s-1}. \tag{52}$$

For $n\kappa = n'\kappa'$, this equation turns into equation (26).

Let us consider first the case $\kappa = \kappa'$. Taking into account that $A_{n'\kappa,n\kappa}^0 = \delta_{n'n}$, we obtain

$$B_{n'\kappa,n\kappa}^0 = \frac{1}{m}(E_{n\kappa} - E_{n'\kappa})(E_{n\kappa}D_{n'\kappa,n\kappa}^1 + mC_{n'\kappa,n\kappa}^1$$
$$+ \alpha Z D_{n'\kappa,n\kappa}^0 - \kappa B_{n'\kappa,n\kappa}^0) + \frac{E_{n\kappa}}{m}\delta_{nn'}. \tag{53}$$

Multiplying this equation with $|n'\kappa\rangle$ and summing over n', we derive

$$|2, 0, \kappa, n\kappa\rangle = \frac{1}{m}(I - |n\kappa\rangle\langle n\kappa|)$$
$$\times (E_{n\kappa}i\sigma_y r + m\sigma_x r + \alpha Z i\sigma_y - \kappa\sigma_z)|n\kappa\rangle, \tag{54}$$

where I is the identity operator. From equations (52) and (54), we can derive the sum $|3, 0, \kappa, n\kappa\rangle$. Then, using equations (45)-(47) and (52), we can calculate all the sums $|i, s, \kappa, n\kappa\rangle$ for $i = 1, 2, 3$ and $s = 0, 1, 2, \dots$. In particular, for the sum $|3, 1, \kappa, n\kappa\rangle$ that occurs in calculations of various corrections to the bound-electron g factor, we find

$$|3, 1, \kappa, n\kappa\rangle = \frac{\kappa}{m^2}(I - |n\kappa\rangle\langle n\kappa|)$$
$$\times \left[\left(E_{n\kappa} - \frac{m}{2\kappa}\right)ri\sigma_y + mr\sigma_x + \alpha Z i\sigma_y - \kappa\sigma_z\right]|n\kappa\rangle. \quad (55)$$

The sums $|4, s, \kappa, n\kappa\rangle$ for $s \neq -1$ can be calculated by employing equation (44). The sum $|4, -1, \kappa, n\kappa\rangle$ is easily derived from the relation

$$D^{-1}_{n'\kappa, n\kappa} = (E_{n\kappa} - E_{n'\kappa})\langle n'\kappa|\ln r|n\kappa\rangle, \quad (56)$$

which is obtained by differentiation of equation (44) with respect to s.

Let us consider now the case $\kappa' \neq \kappa$. From equation (44) we find

$$|3, -1, \kappa', n\kappa\rangle = \frac{1}{\kappa - \kappa'}|n\kappa\rangle. \quad (57)$$

Then, using equations (44)-(47) and (52), we can calculate all the sums $|i, s, \kappa', n\kappa\rangle$ for $i = 1, 2, 3, 4$ and $s = -2, -3, -4, \dots$ (if, of course, the corresponding sum exists). For the sum $|3, -2, \kappa', n\kappa\rangle$ that occurs in calculations of various corrections to the hyperfine splitting, we find

$$|3, -2, \kappa', n\kappa\rangle = \{[1 - (\kappa - \kappa')^2][1 - (\kappa + \kappa')^2] + 4(\alpha Z)^2\}^{-1}$$
$$\times \left[[1 - (\kappa + \kappa')^2]\left(\frac{4\alpha Z m}{\kappa^2 - \kappa'^2} + (\kappa' - \kappa)r^{-1} + r^{-1}\sigma_z\right.\right.$$
$$\left.- \frac{2}{\kappa + \kappa'}(m\sigma_x + E_{n\kappa}i\sigma_y)\right) + \frac{4\alpha Z[(\kappa' + \kappa)m - E_{n\kappa}]}{\kappa - \kappa'}$$
$$\left. + 2\alpha Z[\sigma_x r^{-1} + (\kappa' + \kappa)r^{-1}i\sigma_y]\right]|n\kappa\rangle. \quad (58)$$

For $\kappa = \pm\kappa'$, the corresponding sums can be calculated by taking the limit $\kappa' \to \pm\kappa$. So, taking the limit $\kappa' \to \kappa$ in equation (44), we obtain

$$|3, -1, \kappa, n\kappa\rangle = \frac{\partial}{\partial\kappa}|n\kappa\rangle, \quad (59)$$

where, as in (29), the derivative with respect to κ must be taken at a fixed n_r. Then, the other sums with $\kappa' = \kappa$ and $s = -1, -2, -3, \dots$ can be calculated by using equations (44)-(47) and (52). In particular, we obtain

$$|3, -2, \kappa, n\kappa\rangle = \frac{1}{4(\alpha Z)^2 + (1 - 4\kappa^2)}\left[2\alpha Z\sigma_x r^{-1} + 4\alpha Z\kappa i\sigma_y r^{-1}\right.$$
$$+ (1 - 4\kappa^2)\sigma_z r^{-1} - \frac{(1 - 4\kappa^2)}{\kappa}(E_{n\kappa}i\sigma_y + m\sigma_x)$$
$$\left.- \frac{2(\alpha Z)^3\kappa m}{N^3\gamma}\right]|n\kappa\rangle - \frac{2\alpha Z(2E_{n\kappa} - \frac{m}{\kappa})}{4(\alpha Z)^2 + (1 - 4\kappa^2)}\frac{\partial}{\partial\kappa}|n\kappa\rangle. \quad (60)$$

The case $\kappa' = -\kappa$ can be considered in the same way.

Concluding this section, we give the explicit formulas for $\frac{\partial}{\partial \kappa}|n\kappa\rangle$ for the $1s$ and $2s$ states [14]. For the $1s$ state:

$$\frac{\partial}{\partial \kappa}|n\kappa\rangle = \begin{pmatrix} \tilde{G}_{1s}(r) \\ \tilde{F}_{1s}(r) \end{pmatrix}, \tag{61}$$

where

$$\tilde{G}_{1s}(r) = \frac{k}{\sqrt{1-\gamma}} \exp\left(-t/2\right) t^\gamma \left(\frac{\psi(2\gamma+1)}{\gamma} + (\gamma+1) \right.$$
$$\left. - \frac{1}{2\gamma} - \frac{t}{2} - \frac{1}{\gamma} \ln t \right), \tag{62}$$

$$\tilde{F}_{1s}(r) = -\frac{k}{\sqrt{1+\gamma}} \exp\left(-t/2\right) t^\gamma \left(\frac{\psi(2\gamma+1)}{\gamma} + (\gamma+1) \right.$$
$$\left. + \frac{1}{2\gamma} - \frac{t}{2} - \frac{1}{\gamma} \ln t \right), \tag{63}$$

$$t = \frac{2\alpha Z m r}{N}, \quad k = \frac{(2\alpha Z)^{\frac{3}{2}} m^{\frac{1}{2}}}{2\sqrt{2\Gamma(2\gamma+1)}},$$

$\Gamma(x)$ is the gamma-function, and $\psi(x) = \frac{d}{dx} \ln \Gamma(x)$. For the $2s$ state:

$$\frac{\partial}{\partial \kappa}|n\kappa\rangle = \begin{pmatrix} \tilde{G}_{2s}(r) \\ \tilde{F}_{2s}(r) \end{pmatrix}, \tag{64}$$

where

$$\tilde{G}_{2s}(r) = k' \exp\left(-\frac{t}{2}\right) t^\gamma \frac{\sqrt{2+N}}{N^2-2} \left\{ \frac{t^2}{2(N-1)} \right.$$
$$- \left(\frac{2N^4 - 4N^3 + 5N^2 - 3N + 2}{2N(N-1)^2} + \frac{2\psi(2\gamma+1)}{N-1} \right)t$$
$$+ \frac{N^4 - 2N^3 + N - 2}{2(N-1)} + 2N\psi(2\gamma+1)$$
$$\left. + \frac{2}{N-1} t \ln t - 2N \ln t \right\}, \tag{65}$$

$$\tilde{F}_{2s}(r) = -k' \exp\left(-\frac{t}{2}\right) t^\gamma \frac{\sqrt{2-N}}{N^2-2} \left\{ \frac{t^2}{2(N-1)} \right.$$
$$- \left(\frac{2N^4 - 2N^3 + N^2 + 3N - 2}{2N(N-1)^2} + \frac{2\psi(2\gamma+1)}{N-1} \right)t$$
$$+ \frac{N^5 + 5N^2 - 8N - 4}{2N(N-1)} + 2(N+2)\psi(2\gamma+1)$$
$$\left. + \frac{2}{N-1} t \ln t - 2(N+2) \ln t \right\}, \tag{66}$$

$$t = \frac{2\alpha Z m r}{N}, \quad k' = \frac{\sqrt{\Gamma(2\gamma+2)}}{\Gamma(2\gamma+1)} \frac{1}{\sqrt{8N(N+1)}} \left(\frac{2\alpha Z m}{N} \right)^{\frac{1}{2}}.$$

5 Calculations of the Bound-Electron g Factor and the Hyperfine Splitting in H-Like Atoms

For the last few years a significant progress was achieved in calculations of the bound-electron g factor and the hyperfine splitting in H-like atoms. Formulas (55) and (60) were extensively employed in these calculations.

In [15], a complete αZ-dependence formula for the recoil correction of order m/M to the g factor of an H-like atom was derived. According to this formula, which was confirmed by an independent derivation in [16], the recoil correction is given by $(e < 0)$

$$\Delta g = \frac{1}{\mu_0 m_a} \frac{i}{2\pi M} \int_{-\infty}^{\infty} d\omega \left[\frac{\partial}{\partial \mathcal{H}} \langle \tilde{a} | [p^k - D^k(\omega) + eA_{\text{cl}}^k] \right.$$
$$\left. \times \tilde{G}(\omega + \tilde{E}_a)[p^k - D^k(\omega) + eA_{\text{cl}}^k] | \tilde{a} \rangle \right]_{\mathcal{H}=0} . \tag{67}$$

Here μ_0 is the Bohr magneton, m_a is the angular momentum projection of the state a, $p^k = -i\nabla^k$ is the momentum operator, $\mathbf{A}_{\text{cl}} = [\mathcal{H} \times \mathbf{r}]/2$ is the vector potential of the homogeneous magnetic field $\mathcal{H}$ directed along the z axis, $D^k(\omega) = -4\pi\alpha Z\alpha^l D^{lk}(\omega)$,

$$D^{kl}(\omega, \mathbf{r}) = -\frac{1}{4\pi} \left\{ \frac{\exp(i|\omega|r)}{r} \delta_{kl} + \nabla^k \nabla^l \frac{(\exp(i|\omega|r) - 1)}{\omega^2 r} \right\} \tag{68}$$

is the transverse part of the photon propagator in the Coulomb gauge. The tilde sign indicates that the related quantity (the wave function, the energy, and the Coulomb-Green function $\tilde{G}(\omega) = \sum_{\tilde{n}} |\tilde{n}\rangle\langle\tilde{n}|[\omega - \tilde{E}_n(1-i0)]^{-1}$) must be calculated at the presence of the homogeneous magnetic field $\mathcal{H}$ directed along the z axis. In equation (67) and below, the summation over the repeated indices $(k = 1, 2, 3)$, which enumerate components of the three-dimensional vectors, is implicit. For the practical calculations, this expression is conveniently represented by the sum of the lower-order term and the higher-order term, $\Delta g = \Delta g_{\text{L}} + \Delta g_{\text{H}}$, where

$$\Delta g_{\text{L}} = \frac{1}{\mu_0 m_a} \frac{1}{2M} \left[\frac{\partial}{\partial \mathcal{H}} \langle \tilde{a} | \left\{ \mathbf{p}^2 - \frac{\alpha Z}{r} [(\boldsymbol{\alpha} \cdot \mathbf{p}) + (\boldsymbol{\alpha} \cdot \mathbf{n})(\mathbf{n} \cdot \mathbf{p})] \right\} | \tilde{a} \rangle \right]_{\mathcal{H}=0}$$
$$- \frac{1}{m_a} \frac{m}{M} \langle a | \left([\mathbf{r} \times \mathbf{p}]_z - \frac{\alpha Z}{2r} [\mathbf{r} \times \boldsymbol{\alpha}]_z \right) | a \rangle , \tag{69}$$

$$\Delta g_{\text{H}} = \frac{1}{\mu_0 m_a} \frac{i}{2\pi M} \int_{-\infty}^{\infty} d\omega \left[\frac{\partial}{\partial \mathcal{H}} \langle \tilde{a} | \left(D^k(\omega) - \frac{[p^k, V]}{\omega + i0} \right) \right.$$
$$\left. \times \tilde{G}(\omega + \tilde{E}_a) \left(D^k(\omega) + \frac{[p^k, V]}{\omega + i0} \right) | \tilde{a} \rangle \right]_{\mathcal{H}=0} , \tag{70}$$

where $V(r) = -\alpha Z/r$ is the Coulomb potential induced by the nucleus and $\mathbf{n} = \mathbf{r}/r$. The lower-order term can be calculated analytically by employing formula (55) and the formulas for the average values presented above. Let us consider this calculation in details [15]. According to equation (69), the lower-order term is the sum of two contributions, $\Delta g_{\mathrm{L}} = \Delta g_{\mathrm{L}}^{(1)} + \Delta g_{\mathrm{L}}^{(2)}$. The first contribution is

$$\Delta g_{\mathrm{L}}^{(1)} = \frac{1}{\mu_0 m_a \mathcal{H}} \frac{1}{M} \langle \delta a | \left[\mathbf{p}^2 - \frac{\alpha Z}{r} (\boldsymbol{\alpha} \cdot \mathbf{p} + (\boldsymbol{\alpha} \cdot \mathbf{n})(\mathbf{n} \cdot \mathbf{p})) \right] | a \rangle, \qquad (71)$$

where δa is the first-order correction to the electron wave function due to interaction with the homogeneous magnetic field. Taking into account that $\mathbf{p}^2 = (\boldsymbol{\alpha} \cdot \mathbf{p})^2$ and $(\boldsymbol{\alpha} \cdot \mathbf{p}) = H - \beta m - V$, one easily obtains

$$\langle \delta a | \mathbf{p}^2 | a \rangle = \langle \delta a | (E_a + \beta m - V)(E_a - \beta m - V) | a \rangle$$
$$+ i \langle \delta a | (\boldsymbol{\alpha} \cdot \boldsymbol{\nabla} V) | a \rangle . \qquad (72)$$

The second term in equation (71) can be transformed as (see, e.g., [13])

$$-\langle \delta a | \frac{\alpha Z}{r} [\boldsymbol{\alpha} \cdot \mathbf{p} + (\boldsymbol{\alpha} \cdot \mathbf{n})(\mathbf{n} \cdot \mathbf{p})] | a \rangle = -\langle \delta a | \frac{\alpha Z}{r} \left[2E_a - 2\beta m - 2V \right.$$
$$\left. + \frac{i}{r} (\boldsymbol{\alpha} \cdot \mathbf{n})(\beta \kappa + 1) \right] | a \rangle . \qquad (73)$$

The wave function correction $|\delta a\rangle$ is defined by

$$|\delta a\rangle = \sum_n^{E_n \neq E_a} \frac{|n\rangle \langle n | |e| \boldsymbol{\alpha} \cdot \mathbf{A}_{\mathrm{cl}} | a \rangle}{E_a - E_n} . \qquad (74)$$

Since the operator sandwiched between $|a\rangle$ and $|\delta a\rangle$ in equation for $\Delta g_{\mathrm{L}}^{(1)}$ conserves the angular quantum numbers, we need only that component of $|\delta a\rangle$ which has the same angular quantum numbers as the unperturbed state $|a\rangle$. Using formula (55), we easily find

$$|\delta a\rangle_{\kappa m_a} = \begin{pmatrix} X(r) \Omega_{\kappa m_a}(\mathbf{n}) \\ i Y(r) \Omega_{-\kappa m_a}(\mathbf{n}) \end{pmatrix} , \qquad (75)$$

where

$$X(r) = b_0 \left\{ \left[\frac{2m\kappa - m + 2\kappa E_a}{2m^2} r + \frac{\alpha Z}{m^2} \kappa \right] f(r) + \frac{\kappa - 2\kappa^2}{2m^2} g(r) \right\}, \qquad (76)$$

$$Y(r) = b_0 \left\{ \left[\frac{2m\kappa + m - 2\kappa E_a}{2m^2} r - \frac{\alpha Z}{m^2} \kappa \right] g(r) + \frac{\kappa + 2\kappa^2}{2m^2} f(r) \right\}, \qquad (77)$$

$$b_0 = -\frac{e}{2} \mathcal{H} \frac{\kappa}{j(j+1)} m_a , \qquad (78)$$

$g(r)$ and $f(r)$ are the radial parts of the unperturbed wave function defined above. Integrating over the angular variables in equations (72) and (73), we find

$$\Delta g_{\mathrm{L}}^{(1)} = \frac{\kappa}{j(j+1)} \frac{m}{M} \int_0^\infty dr\, r^2 \Big\{ X(r) g(r) [-2Vm - V^2 + E_a^2 - m^2]$$
$$+ Y(r) f(r) [2Vm - V^2 + E_a^2 - m^2]$$
$$+ [X(r) f(r) + Y(r) g(r)] \frac{\alpha Z}{r^2} \kappa \Big\} . \qquad (79)$$

Substituting expressions (76) and (77) into equation (79), we obtain

$$
\Delta g_{\mathrm{L}}^{(1)} = \frac{\kappa}{j(j+1)} \frac{m}{M} \left\{ \alpha Z \frac{2\kappa E_a - m}{m} C^0 + (\alpha Z)^2 \frac{\kappa}{m} C^{-1} \right.
$$
$$
+ (E_a^2 - m^2) \frac{\kappa}{m} C^1 + \alpha Z \frac{\kappa^2}{2m^2} C^{-2} + (E_a^2 - m^2) \frac{\kappa}{2m^2} A^0
$$
$$
- \alpha Z \frac{\kappa^2}{m} A^{-1} - (\alpha Z)^2 \frac{\kappa}{2m^2} A^{-2} - (E_a^2 - m^2) \frac{\kappa^2}{m^2} B^0
$$
$$
\left. + \alpha Z \frac{3m\kappa - 2\kappa^2 E_a}{2m^2} B^{-1} \right\}. \tag{80}
$$

Using the explicit expressions for the integrals A^s, B^s, and C^s given above, we obtain

$$
\Delta g_{\mathrm{L}}^{(1)} = \frac{m}{M} \frac{\kappa^2}{2j(j+1)} \frac{m^2 - E_a^2}{m^2}. \tag{81}
$$

Consider now the contribution $\Delta g_{\mathrm{L}}^{(2)}$:

$$
\Delta g_{\mathrm{L}}^{(2)} = -\frac{1}{m_a} \frac{m}{M} \left\langle a \left| \left(l_z - \frac{\alpha Z}{2r} [\mathbf{r} \times \boldsymbol{\alpha}]_z \right) \right| a \right\rangle. \tag{82}
$$

Integrating over the angular variables and employing the explicit expressions for the radial integrals derived above, we obtain

$$
\Delta g_{\mathrm{L}}^{(2)} = -\frac{m}{M} \frac{1}{2j(j+1)} \left\{ j(j+1) - \frac{3}{4} + l(l+1) \frac{m + E_a}{2m} \right.
$$
$$
\left. + (2j - l)(2j - l + 1) \frac{m - E_a}{2m} - \kappa^2 \frac{m^2 - E_a^2}{m^2} \right\}. \tag{83}
$$

For the sum $\Delta g_{\mathrm{L}} = \Delta g_{\mathrm{L}}^{(1)} + \Delta g_{\mathrm{L}}^{(2)}$, we find

$$
\Delta g_{\mathrm{L}} = -\frac{m}{M} \frac{2\kappa^2 E_a^2 + \kappa m E_a - m^2}{2m^2 j(j+1)}. \tag{84}
$$

To the two lowest orders in αZ, we have

$$
\Delta g_{\mathrm{L}} = -\frac{m}{M} \frac{1}{j(j+1)} \left[\kappa^2 + \frac{\kappa}{2} - \frac{1}{2} - \left(\kappa^2 + \frac{\kappa}{4} \right) \frac{(\alpha Z)^2}{n^2} \right]. \tag{85}
$$

For the $1s$ state, formula (84) yields

$$
\Delta g_{\mathrm{L}} = \frac{m}{M} (\alpha Z)^2 - \frac{m}{M} \frac{(\alpha Z)^4}{3[1 + \sqrt{1 - (\alpha Z)^2}]^2}. \tag{86}
$$

The first term in the right-hand side of this equation reproduces the result of [17,18]. The higher-order term Δg_{H} was evaluated numerically for the $1s$ state in [19]. Formula (55) was also extensively used in that evaluation.

Let us consider now the derivation of the nuclear-size correction to the g factor of a low-Z H-like atom [20]. To find this correction, we have to evaluate the expression

$$\Delta g = \frac{2}{\mu_0 m_a \mathcal{H}} \sum_n^{n \neq a} \frac{\langle a|\delta V|n\rangle\langle n||e|\boldsymbol{\alpha} \cdot \mathbf{A}_{\mathrm{cl}}|a\rangle}{E_a - E_n}, \tag{87}$$

where δV determines the deviation of the potential from the pure Coulomb one. Integrating over the angular variables, we obtain

$$\Delta g = \frac{2\kappa m}{j(j+1)} \sum_{n'}^{n' \neq n} \frac{\langle n\kappa|\delta V|n'\kappa\rangle\langle n'\kappa|r\sigma_x|n\kappa\rangle}{E_{n\kappa} - E_{n'\kappa}}, \tag{88}$$

where $|n\kappa\rangle$ are the two-component radial wave functions defined above. Substituting expression (55) into this equation, we get

$$\Delta g = \frac{2\kappa^2}{j(j+1)m}\Big\{\langle n\kappa|\delta V\Big[\Big(E_{n\kappa} - \frac{m}{2\kappa}\Big)ri\sigma_y + mr\sigma_x + \alpha Zi\sigma_y - \kappa\sigma_z\Big]|n\kappa\rangle$$
$$- \langle n\kappa|\Big[\Big(E_{n\kappa} - \frac{m}{2\kappa}\Big)ri\sigma_y + mr\sigma_x + \alpha Zi\sigma_y - \kappa\sigma_z\Big]|n\kappa\rangle$$
$$\times \langle n\kappa|\delta V|n\kappa\rangle\Big\}. \tag{89}$$

We assume that the nuclear charge distribution is described by a spherically symmetric density $\rho(\mathbf{r}) = \rho(r)$, which is normalized by the equation

$$\int d\mathbf{r}\, \rho(\mathbf{r}) = 1. \tag{90}$$

The Poisson equation gives

$$\Delta(\delta V)(\mathbf{r}) = 4\pi\alpha Z[\rho(\mathbf{r}) - \delta(\mathbf{r})], \tag{91}$$

where Δ is the Laplacian. When integrated with δV, the radial functions $g(r)$ and $f(r)$ can be approximated by the lowest order term of the expansion in powers of r. It follows that we have to evaluate the integral

$$I = \int_0^\infty dr\, r^2 r^{2\gamma-2}\delta V. \tag{92}$$

Using the identity

$$r^\beta = \frac{1}{(\beta+2)(\beta+3)}\Delta r^{\beta+2} \tag{93}$$

and integrating by parts, we find

$$I = \int_0^\infty dr\, r^2\, \frac{1}{2\gamma(2\gamma+1)}\Delta r^{2\gamma}\,\delta V = \int_0^\infty dr\, r^2\, \frac{1}{2\gamma(2\gamma+1)}\, r^{2\gamma}\,\Delta(\delta V)$$

$$= \frac{4\pi\alpha Z}{2\gamma(2\gamma+1)}\int_0^\infty dr\, r^2\, r^{2\gamma}\, \rho(r) = \frac{\alpha Z}{2\gamma(2\gamma+1)}\, \langle r^{2\gamma}\rangle_{\mathrm{nuc}}, \tag{94}$$

where

$$\langle r^{2\gamma} \rangle_{\text{nuc}} = \int d\mathbf{r} \; r^{2\gamma} \rho(r) \, . \tag{95}$$

For the correction to the g factor we obtain

$$\Delta g = \frac{\kappa^2}{j(j+1)} \frac{\Gamma(2\gamma+1+n_r)2^{2\gamma-1}}{\gamma(2\gamma+1)\Gamma^2(2\gamma+1)n_r!(N-\kappa)N^{2\gamma+2}}$$
$$\times \left[[n_r^2 + (N-\kappa)^2] \left(1 - 2\kappa\frac{E_{n\kappa}}{m} \right) - 2n_r(N-\kappa)\left(\frac{E_{n\kappa}}{m} - 2\kappa \right) \right]$$
$$\times (\alpha Z)^{2\gamma+2} m^{2\gamma} \langle r^{2\gamma} \rangle_{\text{nuc}} \, . \tag{96}$$

For ns-states, which are of particular interest, the expansion of this expression to two lowest orders in αZ yields

$$\Delta g = \frac{8}{3n^3}(\alpha Z)^4 m^2 \langle r^2 \rangle_{\text{nuc}} \left[1 + (\alpha Z)^2 \left(\frac{1}{4} + \frac{12n^2 - n - 9}{4n^2(n+1)} \right. \right.$$
$$\left. \left. + 2\psi(3) - \psi(2+n) - \frac{\langle r^2 \ln(2\alpha Zmr/n) \rangle_{\text{nuc}}}{\langle r^2 \rangle_{\text{nuc}}} \right) \right], \tag{97}$$

where $\psi(x) = \frac{d}{dx} \ln \Gamma(x)$. For the $1s$ state, we have

$$\Delta g = \frac{8}{3}(\alpha Z)^4 m^2 \langle r^2 \rangle_{\text{nuc}} \left[1 + (\alpha Z)^2 \left(2 - C - \frac{\langle r^2 \ln(2\alpha Zmr) \rangle_{\text{nuc}}}{\langle r^2 \rangle_{\text{nuc}}} \right) \right], \tag{98}$$

where $C = 0.57721566490\ldots$ is the Euler constant. In the non-relativistic limit, we find

$$\Delta g = \frac{8}{3n^3}(\alpha Z)^4 m^2 \langle r^2 \rangle_{\text{nuc}} \tag{99}$$

for ns states and

$$\Delta g = \frac{2(n^2-1)}{3n^5}(\alpha Z)^6 m^2 \langle r^2 \rangle_{\text{nuc}} \tag{100}$$

for $np_{\frac{1}{2}}$ states. In the case of the $1s$ state, the expression (99) coincides with the related formula in [21]. A similar derivation for the nuclear-size correction to the hyperfine splitting was performed in [22].

In [23,24], formulas (55) and (60) were employed to evaluate analytically the one-loop vacuum-polarization (VP) corrections to the bound-electron g factor and to the hyperfine splitting in low-Z H-like atoms. These corrections are described by diagrams presented in Fig. 1a–c, where the dashed line ended by a circle indicates the interaction with a magnetic field. According to [23], for the $1s$ state, the VP correction to the g factor calculated in the Uehling approximation

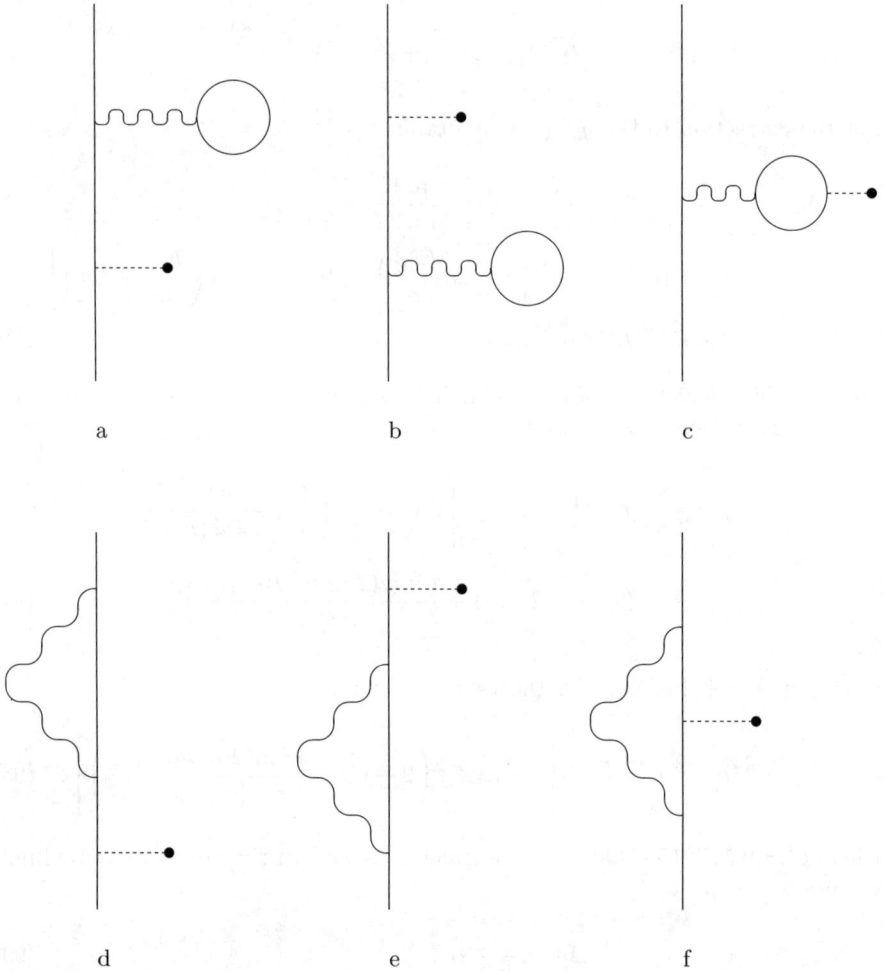

Fig. 1. The first-order QED corrections to the interaction of the electron with a magnetic field

is given by

$$\Delta g = \frac{\alpha}{\pi}\Big[-\frac{16}{15}(\alpha Z)^4 + \frac{5\pi}{9}(\alpha Z)^5$$
$$+\Big(\frac{16}{15}\ln(2\alpha Z) - \frac{2012}{525}\Big)(\alpha Z)^6$$
$$+\Big(-\frac{5\pi}{9}\ln(\alpha Z/2) - \frac{125\pi}{216}\Big)(\alpha Z)^7 + \cdots\Big]. \tag{101}$$

The Uehling correction to the $1s$ hyperfine splitting is [24]

$$\Delta E_{\text{hfs}} = \frac{\alpha}{\pi} E_{\text{F}} \left[\frac{3\pi}{4}(\alpha Z) + \left(\frac{34}{225} - \frac{8}{15} \ln (2\alpha Z) \right)(\alpha Z)^2 \right.$$
$$\left. + \left(\frac{539\pi}{288} - \frac{13\pi}{24} \ln (\alpha Z/2) \right)(\alpha Z)^3 + \cdots \right], \tag{102}$$

where E_{F} is the Fermi energy. In [25,26], the one-loop self-energy corrections to the hyperfine splitting and to the bound-electron g factor (Fig. 1d–f) were calculated numerically to a high accuracy. Formulas (55) and (60) were also employed in these calculations. In [27], formula (55) was used to evaluate the nuclear-polarization effect on the bound-electron g factor.

6   Other Applications of the Virial Relations

Applications of the virial relations are not restricted only by H-like atoms and by the pure Coulomb field. In [14], formulas (58) and (60) were employed to calculate the interelectronic-interaction corrections to the hyperfine splitting in Li-like ions. In [29], virial relations (16)-(19) with $V(r) \neq -\alpha Z/r$ were used to evaluate the recoil correction to the Lamb shift for an extended nucleus. As an example, let us demonstrate how the virial relations with $V(r) \neq -\alpha Z/r$ can be employed to evaluate the nuclear-size correction of the lowest order in mR (R is the nuclear charge radius) to the integral

$$C^{-2} = 2 \int_0^\infty dr \, r^{-2} GF \tag{103}$$

for an ns state, which we denote by $|ns\rangle$. This integral occurs in calculations of the hyperfine splitting. We consider virial relations (16)-(19) for the case $n = n'$, $\kappa = \kappa'$ and for the potential

$$V(r) = -\frac{\alpha Z}{r} + \delta V(r), \tag{104}$$

where δV determines a deviation of the potential from the pure Coulomb one due to the finite-nuclear-size effect. Using notations (20), (21), and (22), where G and F are calculated for potential (104), from equations (16)-(19) we derive

$$A^{-2} - 2B^{-2} = \text{``regular terms''}, \tag{105}$$

$$C^{-2} - 2\langle ns|\sigma_z r^{-1} V|ns\rangle = \text{``regular terms''}, \tag{106}$$

$$2A^{-2} - B^{-2} - 2\langle ns|\sigma_x r^{-1} V|ns\rangle = \text{``regular terms''}, \tag{107}$$

where by "regular terms" we denote terms which have the nuclear-size corrections of order $(mR)^2$ and higher (it means that their integrands have more regular behaviour at $r \to 0$ than the integrand of C^{-2} has). From these equations, we easily obtain

$$C^{-2} = \frac{2}{3 - 4(\alpha Z)^2} \langle ns|(3\sigma_z r^{-1} - 2\alpha Z \sigma_x r^{-1})\delta V|ns\rangle$$
$$+ \text{``regular terms''}. \tag{108}$$

To the lowest order in mR, it follows

$$\delta C^{-2} = \frac{2}{3 - 4(\alpha Z)^2} \langle ns | (3\sigma_z r^{-1} - 2\alpha Z \sigma_x r^{-1}) \delta V | ns \rangle .$$ (109)

An analytical evaluation of this expression to the two lowest orders in αZ yields [28]

$$\delta C^{-2} = \frac{4}{n^3} (\alpha Z)^4 m^3 \langle r \rangle_{\mathrm{nuc}} \left\{ 1 + (\alpha Z)^2 \left[2\psi(3) - \psi(n+1) \right. \right.$$

$$\left. \left. - \frac{\langle r \ln(2\alpha Z mr/n) \rangle_{\mathrm{nuc}}}{\langle r \rangle_{\mathrm{nuc}}} + \frac{8n - 9}{4n^2} + \frac{11}{4} \right] \right\} .$$ (110)

The non-relativistic limit is given by

$$\delta C^{-2} / C^{-2} = -2\alpha Z m \langle r \rangle_{\mathrm{nuc}} .$$ (111)

Formula (111) coincides with the related expression derived in [11] for the sphere model for the nuclear charge distribution, while the relativistic n-independent term in formula (110) differs from the corresponding term that can be derived from the formulas presented in [11]. Since, for the sphere model, the approach developed in [11] provides a more accurate evaluation of the nuclear size correction than the perturbation theory considered here, formula (110) can be improved by replacing the relativistic n-independent term with the corresponding term derived from [11]. As a result, we obtain [22]

$$\delta C^{-2} = \frac{4}{n^3} (\alpha Z)^4 m^3 \langle r \rangle_{\mathrm{nuc}} \left\{ 1 + (\alpha Z)^2 \left[2\psi(3) - \psi(n+1) \right. \right.$$

$$\left. \left. - \frac{\langle r \ln(2\alpha Z mr/n) \rangle_{\mathrm{nuc}}}{\langle r \rangle_{\mathrm{nuc}}} + \frac{8n - 9}{4n^2} + \frac{839}{750} \right] \right\} .$$ (112)

Formulas (110) and (112) differ only by the last constant term.

The virial relations are also helpful for calculations employing finite basis set methods or analytical expressions for the Coulomb-Green function. In particular, they were employed in [30,31] to calculate the nuclear recoil corrections by using the B-spline method for the Dirac equation [32]. In that paper, using the virial relations, the original formulas for the recoil corrections, which contain some integrands with a singular behaviour at $r \to 0$, were expressed in terms of less singular integrals. As a result, the convergence of the numerical procedure for small r was significantly improved. In [8], the virial relations for diagonal matrix elements were used to construct Rayleigh-Schrödinger expansions for eigenvalues of perturbed radial Dirac equations to arbitrary order.

7 Conclusion

In this paper we have considered the derivation of the virial relations for the Dirac equation and their applications for calculations of various physical quantities. It has been demonstrated that the virial relations are a very effective tool for analytical and high-precision numerical calculations of the hyperfine splitting and the bound-electron g factor in H-like ions. They are also useful for calculations employing finite basis set methods and analytical expressions for the Coulomb-Green function.

Acknowledgments

Valuable conversations with D.A. Glazov, V.G. Ivanov, U. Jentschura, S.G. Karshenboim, A.V. Nefiodov, A.V. Volotka, and V.A. Yerokhin are gratefully acknowledged. This work was supported in part by RFBR (Grant No. 01-02-17248), by the program "Russian Universities" (Grant No. UR.01.01.072), and by GSI.

References

1. J. Hirschfelder: J. Chem. Phys. **33**, 1762 (1960)
2. J.H. Epstein, S.T. Epstein: Am. J. Phys. **30**, 266 (1962)
3. S.T. Epstein, J.H. Epstein, B. Kennedy: J. Math. Phys. **8**, 1747 (1967)
4. P. Blanchard: J. Phys. B **7**, 993 (1974)
5. V. Fock: Z. Physik **63**, 855 (1930)
6. S.P. Goldman, G.W. Drake: Phys. Rev. A **25**, 2877 (1982)
7. V.M. Shabaev: Vestn. Leningr. Univ. **4**, 15 (1984)
8. E.R. Vrscay, H. Hamidian: Phys. Lett. A **130**, 141 (1988)
9. V.M. Shabaev: J. Phys. B **24**, 4479 (1991)
10. G.W. Drake, S.P. Goldman: Phys. Rev. A **23**, 2093 (1981)
11. V.M. Shabaev: J. Phys. B **27**, 5825 (1994)
12. V.M. Shabaev: Theor. Math. Phys. **63**, 588 (1985)
13. V.M. Shabaev, A.N. Artemyev: J. Phys. B **27**, 1307 (1994)
14. M.B. Shabaeva, V.M. Shabaev: Phys. Rev. A **52**, 2811 (1995)
15. V.M. Shabaev: Phys. Rev. A **64**, 052104 (2001)
16. A. Yelkhovsky: E-print, hep-ph/0108091 (2001)
17. R.N. Faustov: Phys. Lett. B **33**, 422 (1970); Nuovo Cimento A **69**, 37 (1970)
18. H. Grotch: Phys. Rev. A **2**, 1605 (1970)
19. V.M. Shabaev, V.A. Yerokhin: Phys. Rev. Lett. **88**, 091801 (2002)
20. D.A. Glazov, V.M. Shabaev: Phys. Lett. A **297**, 408 (2002)
21. S.G. Karshenboim: Phys. Lett. A **266**, 380 (2000)
22. A.V. Volotka, V.M. Shabaev, G. Plunien, G. Soff: to be published
23. S.G. Karshenboim, V.G. Ivanov, V.M. Shabaev: JETP **93**, 477 (2001); Can. J. Phys. **79**, 81 (2001)
24. S.G. Karshenboim, V.G. Ivanov, V.M. Shabaev: JETP **90**, 59 (2000); Can. J. Phys. **76**, 503 (1998)
25. V.A. Yerokhin, V.M. Shabaev: Phys. Rev. A **64**, 012506 (2001)
26. V.A. Yerokhin, P. Indelicato, V.M. Shabaev: Phys. Rev. Lett. **89**, 143001 (2002)
27. A.V. Nefiodov, G. Plunien, G. Soff: Phys. Rev. Lett. **89**, 081802 (2002).
28. A.V. Volotka, private communication
29. V.M. Shabaev, A.N. Artemyev, T. Beier, G. Plunien, V.A. Yerokhin, G. Soff: Phys. Rev. A **57**, 4235 (1998); Phys. Scr. T **80**, 493 (1999)
30. A.N. Artemyev, V.M. Shabaev, V.A. Yerokhin: Phys. Rev. A **52**, 1884 (1995); J. Phys. B **28**, 5201 (1995)
31. V.M. Shabaev, A.N. Artemyev, T. Beier, G. Soff: J. Phys. B **31**, L337 (1998)
32. W.R. Johnson, S.A. Blundell, J. Sapirstein: Phys. Rev. A **37**, 307 (1988)

Lamb Shift Experiments
on High-Z One-Electron Systems

Thomas Stöhlker[1,2], Dariusz Banaś[1,3], Heinrich Beyer[1], and
Alexandre Gumberidze[1,2,4]

[1] Gesellschaft für Schwerionenforschung, 64291 Darmstadt, Germany
[2] Institut für Kernphysik, University of Frankfurt, 60486 Frankfurt, Germany
[3] Institute of Physics, Świętokrzyska Academy, 25-406 Kielce, Poland
[4] Tbilisi State University, 380028 Chavchavadze Avenue, Tbilisi, Georgia

Abstract. In this review emphasis is given to x-ray spectroscopic investigations of
the ground-state transitions in H-like uranium by using the intense beams of cooled
heavy ions provided by the storage ring ESR. Such experiments on high-Z ions open
up unique possibilities for the study of the effects of quantum electrodynamics (QED)
in the domain of strong electric field where higher-order corrections can be investigated
which are not accessible for low-Z ions. In particular we will concentrate on the most
current experiment where the deceleration capability of the ESR storage ring was
exploited for the first time for such investigations. From this experiment we deduce
for the ground state of H-like uranium a Lamb Shift value of 468 eV ± 13 eV which
provides us with the most sensitive test of QED for the domain of one-electron ions at
high-Z. Moreover, an overview will be given to the current development of experimental
techniques, aiming at an ultimative test of QED in the high-Z regime, i.e. aiming at a
measurement of the 1s-Lamb-Shift in H-like uranium with an accuracy of 1 eV or even
better.

1 Introduction

Accurate spectroscopic studies of hydrogen-like ions along the isoelectronic sequence is an unique probe for our understanding of relativistic and quantum
electrodynamic effects in the atomic structure. It was for atomic hydrogen where
Lamb and Retherford discovered the small difference between the binding energies of the $2s_{1/2}$ and $2p_{1/2}$ states [1], the so-called Lamb shift, caused essentially
by the effects of quantum-electrodynamics (QED). The Lamb shift is defined as
the difference between the real binding energy and the Dirac-Coulomb energy
for a point-like nucleus. It leads to a reduced binding energy for all atomic levels
but is largest for s-states. This is due to the large overlap of their wave functions
with the nucleus. In general the Lamb Shift can be presented by the following
equation [2,3]:

$$L = \frac{\alpha}{\pi} \frac{(Z\alpha)^4}{n^3} \cdot F(Z\alpha) \cdot m_0 c^2 \qquad (1)$$

where α is the fine-structure constant, n is the principal quantum number, $m_0 c^2$,
the electron rest mass, and $F(Z\alpha)$ is a slowly varying function of Z respectively,
This function considers all the QED corrections and includes in addition the

Fig. 1. Feynman diagrams for the self energy (left) and the vacuum polarization (right) of order α for a bound electron. To denote the bound state, the electron lines in the Feynman diagrams are doubled. On the left, an electron emits and reabsorbs a virtual photon. The loop on the right represents a virtual electron-positron pair (also in the field of the nucleus) that is created and re-annihilated

effect caused by the finite size of the nucleus. Since the Lamb shift scales approximately with Z^4/n^3, all these corrections are largest for the ground-state and for the strong fields at high-Z.

Within the theory of QED, the Lamb Shift is explained by the interaction of the electron with its own radiation field and the main contributions for the Lamb shift arise from the following effects:

(a) The self energy, i.e. the emission and reabsorption of a virtual photon (at low-Z the self energy, gives the most important Lamb shift correction). The corresponding Feynman diagram is depicted at the left side of Fig. 1.

(b) The vacuum polarization of the electron bound in the external field of the nucleus, i.e. the virtual creation and annihilation of an electron-positron pair (with increasing nuclear charge the influence of the vacuum polarization increases continuously). The corresponding Feynman diagram is depicted at the right side of Fig. 1.

For illustration, in Fig. 2 the contributions of the self energy, vacuum polarization, and of the finite nuclear size to the Lamb Shift in hydrogen-like atoms are given separately as a function of the nuclear charge.

For light one-electron systems such as atomic hydrogen, the theory of QED is now well confirmed with extraordinary precision [4–6]. However, for a test of the higher order terms, the strong Coulomb fields of the heaviest species, such as H-like uranium, are required. At high-Z, the influence of the higher-order contributions becomes so important that the radiative corrections must be calculated to all orders of αZ [7–9]. For such ions, one basic experimental approach for the investigation of the effects of QED is a precise determination of the x-ray energies emitted by transitions to the ground-state of the ion (for the origin of the Lyman-α transitions which are of particular relevance, see Fig. 2). Here, the goal of the experiments is to achieve a precision which not only tests the higher order contributions in αZ but also probes QED corrections which are beyond the one-photon exchange corrections, such as the two-photon exchange diagrams, i.e. α^2 contributions. The relevant Feynman diagrams are depicted in Fig. 3.

In Table 1 all theoretical values for the individual Lamb-shift corrections of the 1s-state in U^{91+} are given [10]. The theoretical binding energies for the first

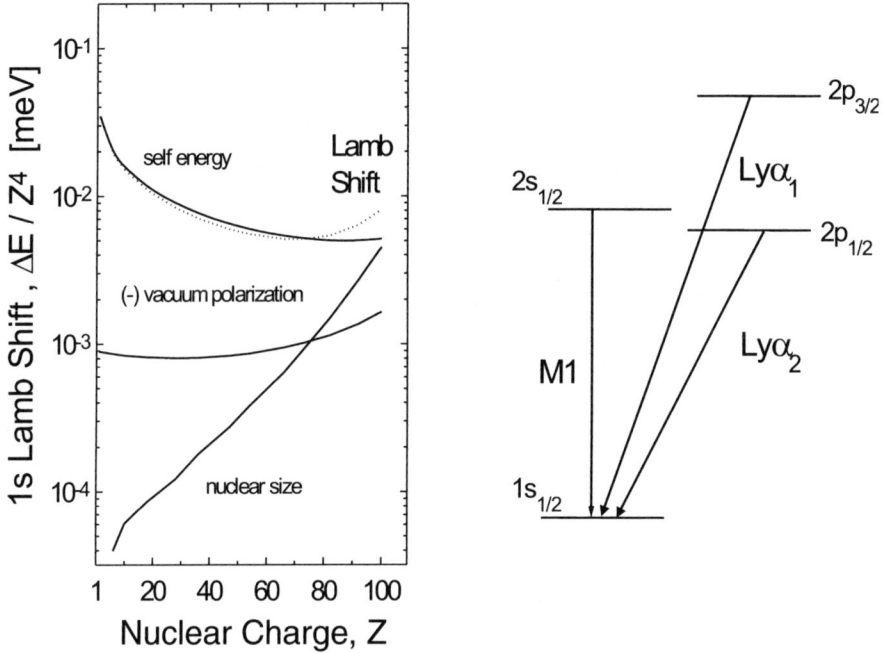

Fig. 2. The various contributions to the ground state Lamb shift in hydrogen-like ions as a function of the nuclear charge, according to Ref. [3]. In addition a schematic presentation of the origin of the Lyman-α transitions is given

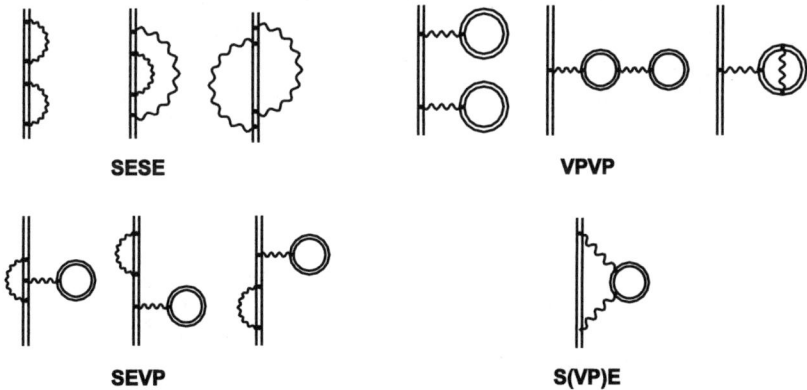

Fig. 3. Feynman diagrams for the QED-corrections of order α^2 for H-like ions (see e.g. [10]

Table 1. The contributions to the 1s binding energy in H-like uranium [10]. The corresponding binding energy for the $2s_{1/2}$ and the $2p_{1/2}$ state amounts to -34127.78 eV and -34204.14 eV, respectively [11]. For the $2p_{3/2}$ level the binding energy is calculated to -29640.99 [12] whereby the Lamb shift corrections amounts to $+8.8$ eV

Dirac Value	-132279.96 eV
Finite nuclear size	198.81 eV
Nuclear Recoil	0.46 eV
Nuclear Polarization	-0.20 eV
VP (Fig. 1)	-88.60 eV
SE (Fig. 1)	355.05 eV
SESE (Fig. 3)	-1.87 eV
VPVP (Fig. 3)	-0.97 eV
SEVP (Fig. 3)	1.14 eV
S(VP)E (Fig. 3)	0.13 eV
Lamb shift	463.95$\pm$0.5 eV
Binding energy	-131816.01 $\pm$0.5 eV

Table 2. Theoretical energies for the individual L→K transitions in H-like uranium as derived from Table 1

	Lyman-α_1	Lyman-α_2	**M1**
	$2p_{3/2} \to 1s_{1/2}$	$2p_{1/2} \to 1s_{1/2}$	$2s_{1/2} \to 1s_{1/2}$
Energy (eV)	102 175.0	97 611.9	97 688.2

excited states are quoted in the table caption as well. The resulting $L \to K$ transition energies are listed separately in Table 2.

The ESR storage ring with its brilliant beams of cooled heavy-ions provides unique conditions for this kind of precision investigations. This will be shown in this review by discussing the experiments which were conducted up to now at the ESR for the particular case for hydrogen-like uranium [13–19] (for a review see also Ref. [20–22]).

2   The Storage Ring ESR

The production and cooling of intense beams of fully stripped ions, as introduced by heavy-ion storage rings and in particular by the storage ring ESR, constitutes an important step for accurate precision spectroscopy of atomic transitions in the realm for high-Z systems.

Figure 4 shows a schematic sketch of the ESR storage ring (circumference of 108 m, magnetic rigidity of 10 Tm) and its main components such as the electron cooler device, the internal gasjet target, and the rf-cavities. For the experiments, highly-charged ions are accelerated in the heavy ion synchrotron (SIS) to the final energies of typically 200 to 400 MeV/u and are extracted into the transfer line towards the ESR. In the transfer line the ions pass through a thick Cu stripper foils. From the emerging charge state distributions, the fraction of bare ions is magnetically separated and injected into the storage ring. Note

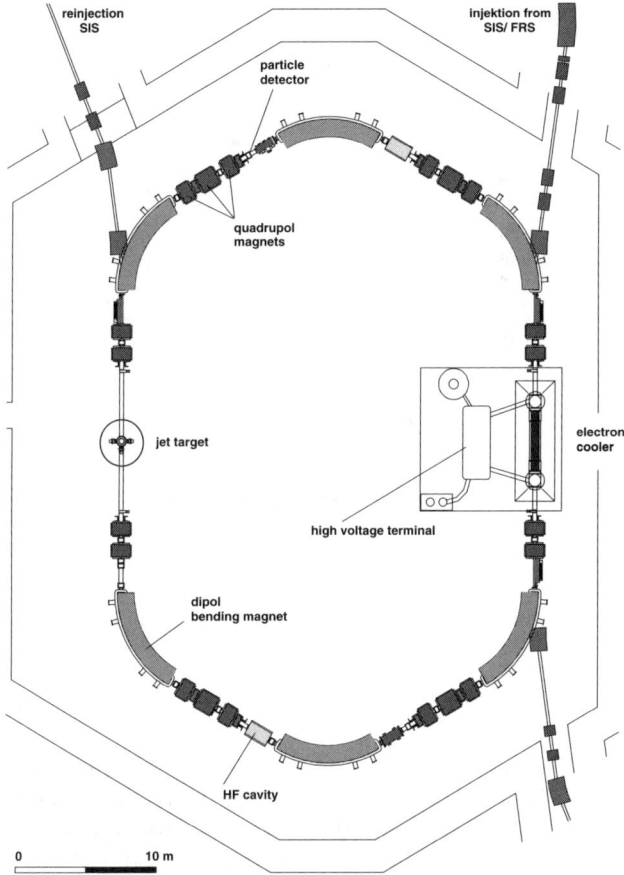

Fig. 4. The experimental storage ring ESR

that for the case of uranium ions where the K-shell binding energy amounts to ≈ 130 keV, a beam energy of at least 300 MeV/u is required in order to produce bare ions with sufficient intensity, a beam energy which corresponds approximately to $\beta = 0.6$, where β denotes the ion velocity in units of the speed of light. In the storage ring, the injected hot ion beam with a typical emittance of about 5 π mm mrad is very efficiently cooled by Coulomb interaction with the co-moving electrons in the 2 m long electron cooler section. For this purpose electron currents of typically 100 to 300 mA are applied. This cooling technique leads to an emittance of the stored beam of less than 0.1 π mm mrad and to a small beam size with a typical diameter of less than 5 mm. In particular, electron cooling guarantees for a well defined constant beam velocity which is generally of the order of $\Delta\beta/\beta \approx 10^{-4}$. It also reduces the relative longitudinal momentum spread of the injected ion beam of $\Delta p/p \approx 10^{-3}$ to about 10^{-5} (for a detailed discussion of the electron cooling technique we refer to [23]). However, both the transverse emittance and the relative momentum spread of the stored beam depend on the number of stored ions and the applied cooler current [23].

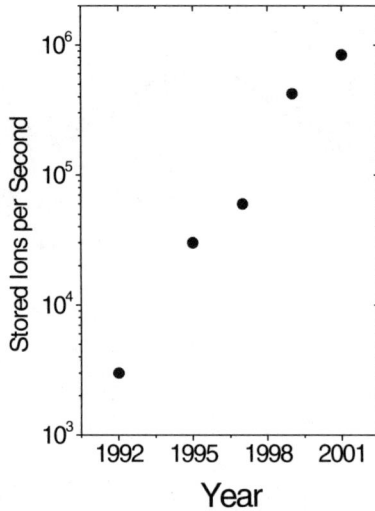

Fig. 5. Effective number of ions stored in the experimental storage ring ESR

Within the last years the maximum possible number of stored ions was improved significantly (see Fig. 5). For high-Z ions, e.g. uranium, up to 10^8 ions can be stored routinely. This number is still below the upper limit of particles which can be stored in principle. These limits are due to the space charge potential of the stored ion beams and restrict the number of stored ions e.g. for the case of bare uranium at 556 MeV/u to 9.3×10^9 and at 50 MeV/u to 4.4×10^8, respectively [24].

The ESR is up to now the only storage ring equipped with a gasjet target. Here, various gas targets such as CH_4, N_2, Ar or even heavier targets can be used with areal densities of about 10^{12} particles/cm^2 and a diameter of about 5 mm. These areal densities are rather low compared to those of solid targets used in standard single pass experiments at accelerators (about 10^{18} particles/cm^2). However, the low gas density is compensated by the high revolution frequency of the circulating ions in the ring of about 10^6 1/s .

3 X-Ray Spectroscopy at the ESR

The individual 1s Lamb shift experiments conducted up to now applied quite different techniques and methods. The underlying principle of all Lamb shift investigations at high-Z, however, is the same for all experiments and it can be summarized as follows:

- production of the bare ion species,
- storing and cooling inside the storage ring,

- population of excited levels via electron capture,
- detection of Lyman-α radiation,
- determination of the line centroids,
- transformation of the results into the emitter frame,
- and finally, from the difference between the measured transition energies and the predictions of the Dirac theory for a point like nucleus, one obtains the desired experimental Lamb shift values.

3.1   The Experimental Challenge: Doppler Corrections

Both the gasjet-target and the electron cooler can be applied for an intense production of characteristic Lyman-α radiation of the circulating high-Z ions. At the gasjet target, capture of bound target electrons into the fast moving, bare projectiles populates excited levels of H-like ions and finally results in the emission of Lyman photons. At the electron cooler side, the free electrons are captured via radiative recombination (i.e. the time reversed photo ionization process) into the bare ions, populating also excited levels of the H-like species formed by the capture process. By cascades, all such events lead to Lyman photon emission. Although the ESR provides brilliant, monochromatic beams, the main problem encountered is still caused by the uncertainties introduced by the Doppler shift because the x-rays are emitted by ions moving with velocities of about 60% of the speed of light. In order to derive the Lyman-α transition energy in the emitter frame, the transition energy measured in the laboratory system must be corrected for the relativistic Doppler shift given by

$$E = E_{\text{lab}} \cdot \gamma \cdot (1 - \beta \cos\theta_{\text{lab}}). \tag{2}$$

Here, E and E_{lab} are the x-ray energies in the emitter system and in the laboratory frame, respectively. θ_{lab} denotes the laboratory observation angle and γ is the relativistic Lorentz factor. In Fig. 6a,b the ratio E_{lab}/E is plotted as a function of observation angle for two different beam energies. The final uncertainty of the x-ray energy in the emitter frame is determined by the uncertainties in the absolute value of β and of the observation angle θ_{lab}. The influence of the latter on the final result depends crucially on the beam velocity and the observation angle chosen. This can easily be seen from the derivative of (2) given by

$$\left(\frac{\Delta E}{E}\right)^2 = \left(\frac{\beta \sin\theta_{\text{lab}}}{1 - \beta\cos\theta_{\text{lab}}}\Delta\theta_{\text{lab}}\right)^2 + \left(\gamma^2 \frac{\cos\theta_{\text{lab}} - \beta}{1 - \beta\cos\theta_{\text{lab}}}\Delta\beta\right)^2 + \left(\frac{\Delta E_{\text{lab}}}{E_{\text{lab}}}\right)^2 \tag{3}$$

For instance, due to the $\sin\theta_{\text{lab}}$ term, the uncertainty in $\Delta\theta_{\text{lab}}$ does not affect the final result at observation angles close to $0°$ and $180°$. Here, the error due to $\Delta\beta$ is largest. Also, by choosing $\beta = \cos\theta_{\text{lab}}$ the uncertainty caused by $\Delta\beta$ can be minimized but now the uncertainty introduced by $\Delta\theta_{\text{lab}}$ is maximal (see Fig. 6c,d). In practice a velocity-sensitive measurement at the electron cooler and an angular-sensitive geometry at the gas-jet target were realized in the experiments conducted up to now at the ESR. This way absolute observation angles are either not critical or they are spectroscopically determined by using several

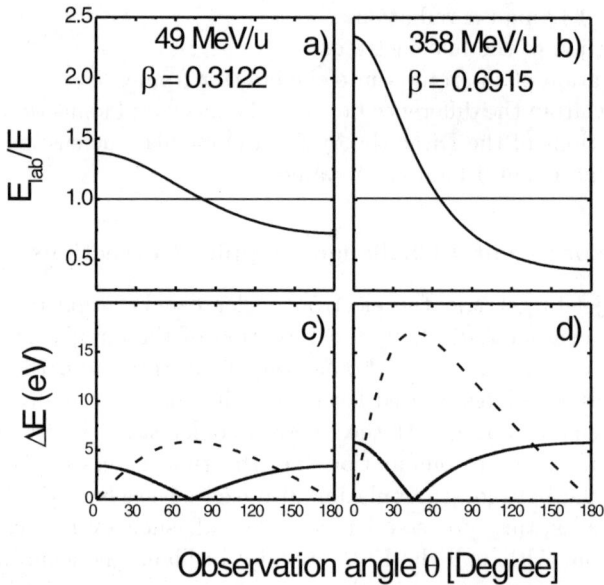

Fig. 6. a,b The ratio E_{lab}/E as a function of the observation angle for two beam energies; **c,d** the uncertainty for the Lyman-α_1 transition energy in U^{91+} for the emitter frame. Dotted line: assumed uncertainty of the β-value of $\Delta\beta = 3 \cdot 10^{-5}$; solid line: assumed uncertainty in the observation angle θ of $\Delta\beta = 0.01°$

detectors viewing the same interaction zone. For completeness it is important to note that $\Delta\theta_{lab}$ and $\Delta\beta$ can also be interpreted as widths. Hence (3) describes also the Doppler width observed in the laboratory frame.

3.2   Experiments at the Electron Cooler

The experimental setup for the measurements of x-ray radiation at the electron cooler device is shown in Fig. 7 [16,17]. At the electron cooler, the ion-beam/electron-beam interaction region is viewed by a solid state Ge(i) detector at an observation angle close to 0.55°, i.e. close to 0°, where a slight uncertainty in the observation angle does not affect the final precision (see above). The detector is mounted 4.2 m downstream of the midpoint of the 2 m long straight electron cooler section which results in a solid angle of $\Delta\Omega/\Omega = 4 \times 10^{-5}$. The x-rays are produced by electron capture into the bare projectiles and recorded in coincidence with the down-charged ions. For this purpose a position sensitive multi-wire detector is installed behind the first dipole magnet, located downstream from the electron cooler section. By using this experimental setup, Lamb shift experiments have been performed for H-like Au^{78+} and U^{91+} at specific beam energies of 298 MeV/u and 321 MeV/u, respectively [15–17].

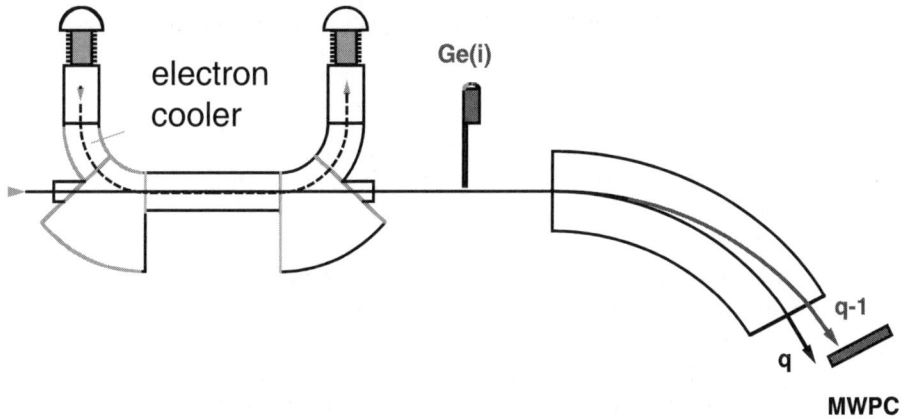

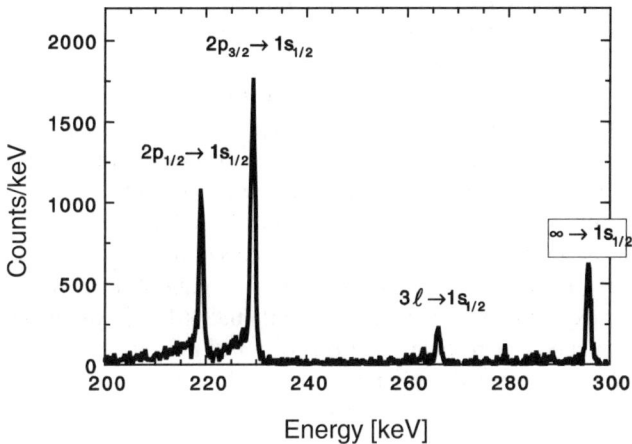

Fig. 8. Coincident x-ray spectrum of U^{91+} measured at the electron cooler for an ion-beam energy of 321 MeV/u at $0°$ observation angle [16,17]

As an example, the x-ray spectrum for initially bare uranium ions undergoing electron capture in the cooler is shown in Fig. 8. The spectrum is almost background free, because it was recorded in coincidence with down charged U^{91+} ions. Moreover, due to the observation angle of approximately $0°$, the characteristic Lyman-α transitions, with an x-ray energy of about 100 keV in the emitter frame, are strongly blue shifted and appear at energies close to 220 keV. Also an x-ray line close to about 300 keV is clearly visible in the spectrum caused by radiative recombination transitions into the vacant $1s$-shell of the bare uranium ions. Note, that the radiative recombination (RR) process at low relative velocities populates predominantly high-n, l states. As a consequence it can be shown that the $2p_{1/2} \rightarrow 1s_{1/2}$ ground-state transition is practically not contaminated by the $2s_{1/2} \rightarrow 1s_{1/2}$ M1 decay. Moreover, the fact that capture at the cooler

Table 3. Experimental result for the 1s Lamb Shift in H-like uranium as obtained from the electron cooler experiment at the ESR storage ring [15–17]. All values are given in eV

	1s-Binding Energy	1s-Lamb Shift	Statistical Uncertainty	$\Delta\beta$
Z=92	131 810 ±16	470±16	±6.9	±9.1

populates high n,l states explains also the distinctive tails on the low-energy side of the Lyman transitions which are observed in the x-ray spectrum. The cascades following electron capture into highly excited levels may lead to a delayed Lyman emission, which then takes place within the 3m long distance between the end of the electron cooler and the Ge(i) detector. Such events are measured at observation angles between 0.8° and 9° which gives rise to an appreciable Doppler shift towards lower x-ray energies [15,16]. As these line profiles can be modelled precisely by cascade feeding calculations, no reduction in the final accuracy is caused by these delayed transitions. Again it is important to point out that the investigations at the cooler were conducted at almost 0° observation angle. It leads to a very small Doppler broadening and makes the experiment insensitive to slight uncertainties in the observation angle. However, the experiments are sensitive to $\Delta\beta$, as discussed above. In order to determine precisely the absolute beam velocity three different methods were employed: (1) the measurement of the revolution frequency, (2) the measurement of the high-voltage of the electron cooler, (3) the determination of energy differences between the lines appearing in the x-ray spectrum. All three methods led to a consistent result. For the case of the uranium experiment a β-value of $\beta = 0.66884(3)$ was deduced by the high voltage which turned out to be almost one order of magnitude more precise than the values obtained from the other two methods [15–17]. The results obtained at the electron cooler for the Lamb shift of U^{91+} are summarized in Table 3. In addition the error contributions due to the uncertainties in beam velocity and counting statistics are given separately in the table.

Very recently, July 2002 [25], a new experiment was conducted at the electron cooler. Here the deceleration technique of the ESR storage ring was exploited for the first time in an electron cooler experiment by decelerating U^{92+} ions down to 20 MeV/u. As a consequence all uncertainties associated with Doppler corrections are strongly reduced compared to high-energy beams. Also, for decelerated ions the bremsstrahlung intensity caused by the cooler electrons is strongly reduced (due to the comparably small cooler voltage of 11 kV and an electron beam current of 50 mA). Consequently, very clean conditions for x-ray spectroscopy are present at the cooler section. This is depicted in Fig. 9a, where a preliminary calibrated, coincident x-ray spectrum is plotted as observed for initially bare uranium ions at an energy of 20.8 MeV/u. Note, the strongly reduced bremsstrahlung intensity allowed us for the very first time to observe even RR transitions into the L-shell (L-RR, see Fig. 9) as well as Balmer radiation. All these transitions are located at the low-energy part of the spectrum. A further important aspect of our study is that due the low β-value of 0.21 and the exper-

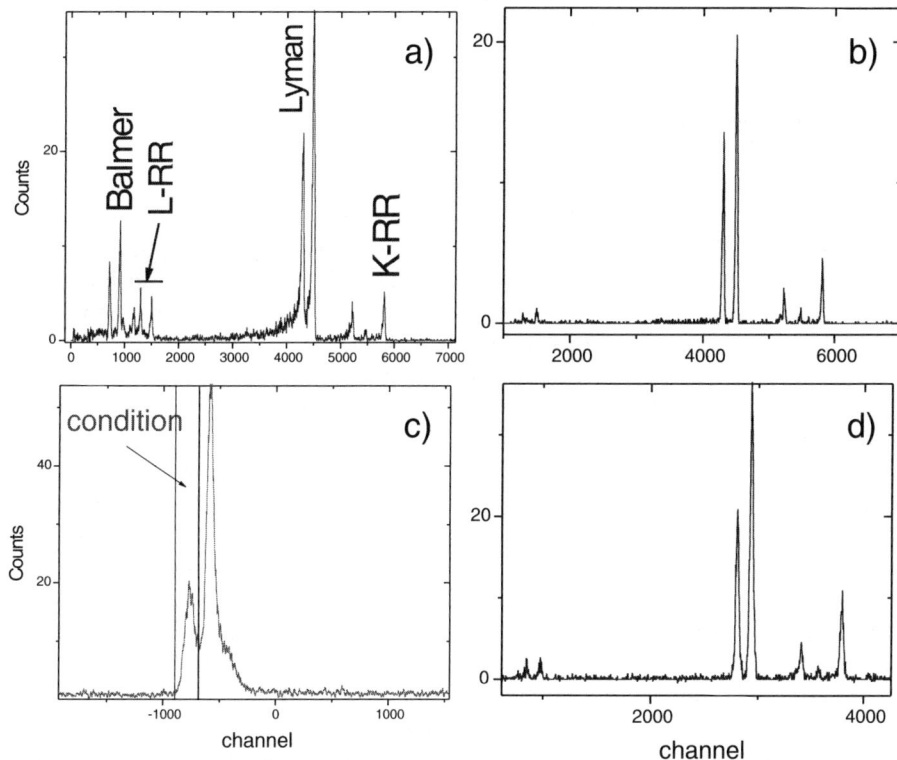

Fig. 9. a Preliminary result for a coincident x-ray spectrum of U^{91+} measured at the electron cooler for an ion-beam energy of 20.8 MeV/u at $0°$ observation angle [25]. b the same as a, but accumulated in addition with the condition on the coincidence time spectrum, c Coincidence time spectrum, d Coincident x-ray spectrum recorded at an observation angle of $180°$

imental time resolution of about 20 ns, photon events which occurred inside the cooler section can be distinguished from events where the emission took place just in front of the x-ray detector (see Fig. 9c). For the latter x-ray events, the set-up possesses a comparably large solid angle and the photon energy appears markedly shifted leading to the low-energy tails of the Lyman radiation (as it was already mentioned above). In Fig. 9b we depict the photon spectra which were accumulated by using a time condition which excludes most of the cascade contributions leading to the low-energy tails. Indeed, as observed in Fig. 9b, the time condition used eliminates almost completely the low energy tails and reduces also the intensity of the Lyman transitions but leads the K-RR lines practically unaffected. In Fig. 9d we depict a photon spectrum recorded at $180°$ observation angle. Note that there are no tails present for the Lyman lines. This result is in agreement with the assumption that the tails are generated by delayed transitions which take place out of the electron cooler and just in front of the x-ray detector. Currently the data evaluation is in progress [25].

3.3 Experiments at the Internal Gasjet Target

The experimental arrangement at the ESR gasjet target is illustrated in Fig. 10. It shows the ion/beam target interaction zone surrounded by four Ge(i) x-ray detectors mounted at observation angles of 48°, 90°, and 132° with respect to the predicted ion-optical beam axis. This setup allows for an intrinsic control of the beam/target/x-ray detector geometry by the simultaneous use of various x-ray detector devices at different observation angles [13,14,19,26].

The two detectors at 48° are installed symmetrically on opposite sides of the reaction chamber. This observation angle is, at high β-values, close to the magic angle of $\theta_{\mathrm{lab}} = \cos(\beta)$ where the experiment is insensitive to uncertainties in the beam velocity. However, here the Doppler broadening and the uncertainties introduced by the uncertainty in observation angle are largest (see above). Therefore, one of these 48° detectors is a conventional solid state detector equipped with an x-ray collimator in order to confine the angular acceptance, thus reducing the Doppler broadening. The other detector consists of seven equidistant, parallel segments each furnished with a separate electronic readout. They deliver seven independent x-ray spectra. The resulting sum spectrum combines the advantage of the large solid angle with a narrow Doppler width of one segment.

At 90° a similar segmented detector is used whereas at 132° a conventional Ge(i) detector is installed. The exact geometry of the whole detector arrangement is measured by laser assisted trigonometry. Knowing precisely the relative angles between all the detectors the individual position of each x-ray detector can be determined by assuming that the origin of the radiation is the same for all detectors. This procedure is required, as the absolute position of the gasjet/beam interaction zone is not precisely know [13].

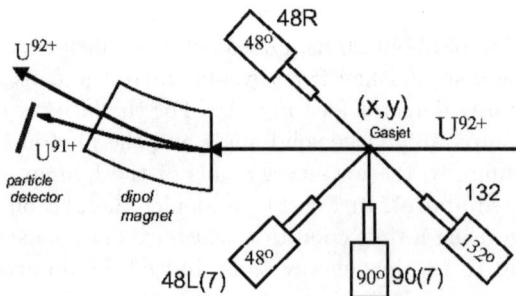

Fig. 10. Experimental arrangement at the ESR gasjet target. X-rays are measured in coincidence with the down-charged U^{91+} projectiles detected in the particle detector [19]

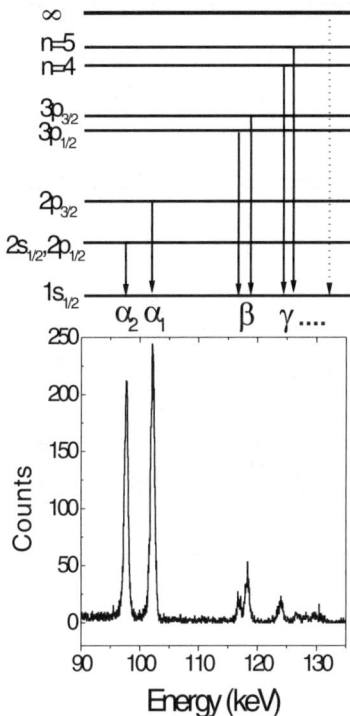

Fig. 11. Lyman spectrum measured by one segment of the $48°$ strip detector in coincidence with one electron capture for decelerated uranium ions at 68 MeV/u ($U^{92+} \to N_2$). In the level diagram the origin of the various Lyman transitions are shown [19]

The x-ray emission produced via electron pickup from the gasjet particles (usually a N_2 gasjet target is used) into the fast moving bare projectiles is registered by the x-ray detector array in coincidence with projectiles having captured one electron.

For the latter purpose a fast plastic scintillator counter for detection of the down-charged U^{91+} ions is used which is located behind the dipole magnet. A typical coincident x-ray spectrum as observed for $U^{92+} \to N_2$ collisions at 358 MeV/u is shown in Fig. 12a. Besides the well resolved Lyman-α transitions observed, the spectrum is entirely dominated by radiative electron capture (REC) transitions into the projectile, a process very similar to radiative recombination except that bound target electrons are captured instead of free electrons. In first investigations conducted at the gasjet target, the discussed experimental technique was applied for U^{92+} ions colliding with the N_2 gas-jet particles at 294.7 MeV/u [13]. At the time this former experiment was performed the uncertainty in the determination of the beam velocity was rather large ($\Delta\beta/\beta \approx 4 \times 10^{-4}$).

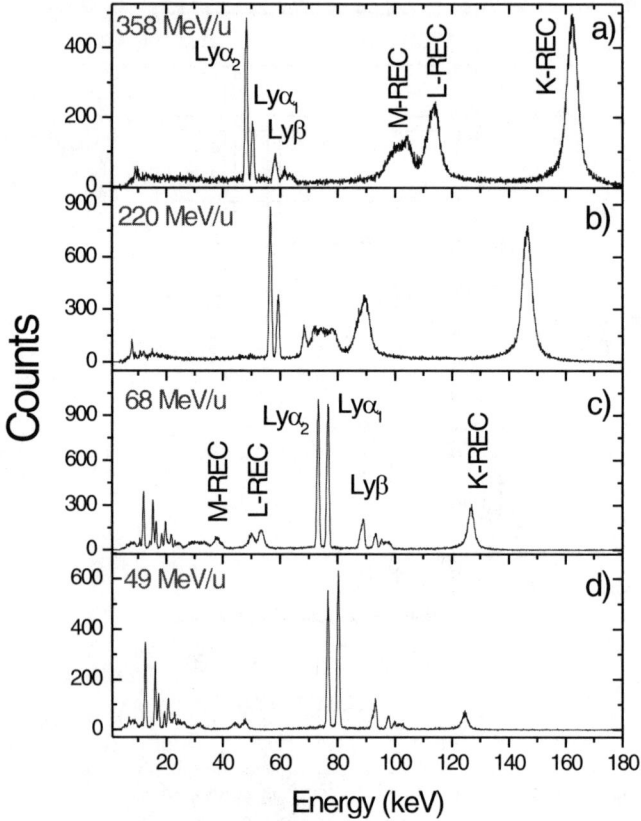

Fig. 12. X-ray spectra (projectile system) of H-like uranium measured at **a** 358 MeV/u, **b** 220 Mev/u, **c** 68 MeV/u, **d** 49 MeV/u [27]

Moreover, only up to 10^7 ions could be stored in the ring and only one segmented detector at 48° and the conventional detector at 132° were used. The accuracy of the final result of 429 eV $\pm$ 63 eV [13] was determined essentially by counting statistics. Meanwhile much higher beam currents are available at the ESR, as demonstrated by the experiments performed at the electron cooler. In order to overcome the drawbacks associated with fast moving sources, this experiment was carried out at various beam energies [19,26].

In particular, the deceleration mode of the ESR storage ring was applied for the first time. It provides bare uranium ions at moderate energies as low as 49 MeV/u which reduces strongly the uncertainties associated with Lorentz transformation to the emitter system. Bare uranium ions with an energy of 360 MeV/u delivered from the SIS were injected into the ESR storage ring. After finishing the stacking procedure, which was still performed at the energy of 360 MeV/u, the coasting DC-beam was rebunched and decelerated by simultaneously ramping down the magnetic fields to the chosen final beam energies of 220, 68, and 49 MeV/u. Subsequently, electron cooling at the new energy was switched on. By applying this procedure up to 2×10^7 ions could be decelerated with losses below 20%. The potential of the deceleration capabilities of the ESR is illustrated by the spectra of hydrogen-like uranium also shown in Fig. 12. Compared to the high beam energy of 358 MeV/u, the x-ray spectra recorded for the decelerated ions (68 MeV/u) provide an abundant yield of different characteristic projectile transitions. At low beam energies, where non-radiative electron capture (NRC) dominates, electron capture populates predominately highly excited levels. Such capture events lead through cascades to Balmer ($n = 3, 4, ... \rightarrow n = 2$) as well as to Lyman transitions ($n = 2, 3, 4, ... \rightarrow n = 1$). Obviously (see Fig. 11), the production of characteristic projectile x-rays is now much more efficient than at high energies where REC to the ground state prevails.

In order to elucidate in more detail the advantages of the low collision velocity, in Fig. 11 the transformed Lyman spectrum (emitter system) is shown recorded by one segment of the strip detector mounted at 48° (68 MeV/u, $U^{92+} \rightarrow N_2$ collisions). The most striking feature of this spectrum is the observed splitting of the Lyman-β line into two components ($3p_{1/2} \rightarrow 1s_{1/2}, 3p_{3/2} \rightarrow 1s_{1/2}$) which makes this experiment also sensitive to the M-shell fine structure splitting. The gain in resolving power is a consequence of the strongly reduced Doppler broadening ($\Delta E_{\text{lab}}/E_{\text{lab}} = 5 \times 10^{-3}$) caused by the small angle acceptance of $\pm$ 0.320 of this particular detector as well as by the low β-value. Finally it has to be mentioned that due to the large reaction cross-section at low energies, the stored ion beams are used up very efficiently.

Knowing precisely the relative angles between all the detectors the individual position of each x-ray detector can be determined via (2) by assuming that the origin of the radiation is the same for all detectors used. This procedure is required, as the absolute position of the gasjet/beam interaction zone is not precisely known. Therefore, the whole detector arrangement, i.e. the detector positions relative to each other, were measured within a cartesian coordinate system by applying laser assisted trigonometry. This measurement was conducted just before and directly after the experiment. From a comparison of both measurements, a mean angular stability of the set-up of $\pm 0.015°$ was deduced. Besides the mechanical stability, the control over possible electronic drifts was most important. For this purpose, all detectors were calibrated frequently during the experiment by using [169]Yb and [182]Ta standard calibration lines. In total 30 calibration runs were conducted during a total beam time of 15 days. Finally, the centroid energies of the groundstate transition lines of each detector were fitted with Gaussian distributions (germanium detector response functions) [29]. Since the line shapes might be influenced by the Doppler effect, the appropriateness of the applied fit procedure was checked by Monte Carlo simulations [30]. Typically, an accuracy for the determination of the centroid energies of 10 to 20 eV was achieved. The latter turned out to be limited by calibration uncertainties rather than by counting statistics.

The centroids of the Lyman-α transitions in the projectile frame were determined separately for the two beam energies of 68 and 49 MeV/u. For this purpose

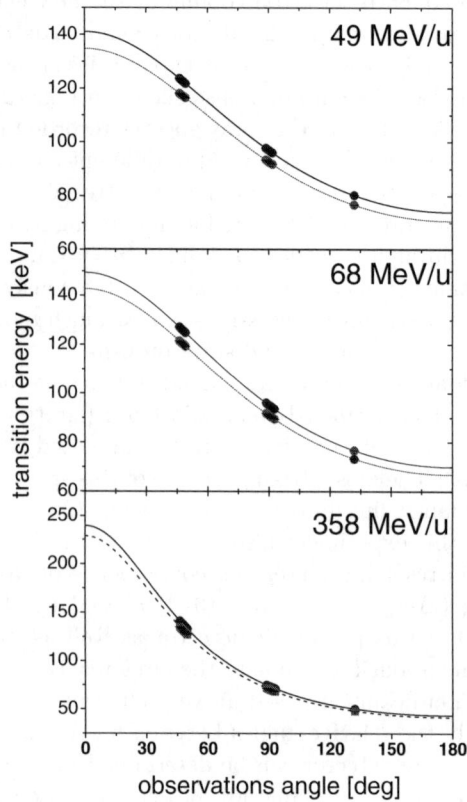

Fig. 13. Experimental Lyman-α transition energies as function of observation angle (Lyman-α_1:solid line; Lyman-α_2+**M1**: dashed line. The lines refer to the result of a least square minimizing procedure, considering the the measured centroid energies as well as known relative angles between the individual detectors [28]

an uncertainty minimizing procedure based on (1) was applied, by considering the known positions of each detector/detector-segment. Within this procedure, the origin (x,y-coordinates) of the ion-beam/target interaction point was spectroscopically determined by considering all Lyman-α centroid energies of the individual segments at the various observation angles. This is illustrated in Fig. 13. Also, the uncertainty in the knowledge of the absolute beam velocity has to be considered. As established recently, the latter is assumed to be caused by a systematic uncertainty in the cooler voltage of 10 V. For the decelerated ions used in our experiment where cooler voltages of 32277 V (at 68 MeV/u) and 26822 V (at 49 MeV/u) were applied this assumption is quite conservative [31]. However, the latter only slightly affects the overall accuracy of the experimental results.

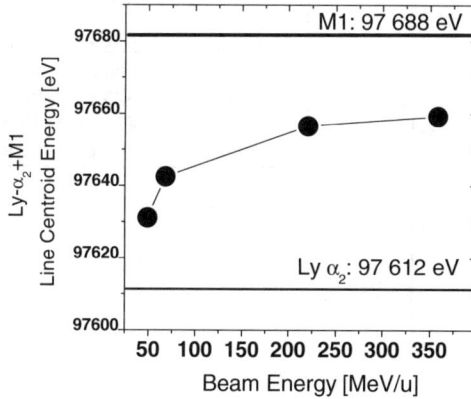

Fig. 14. Line centroids of the Lyman-α_2+M1 x-ray line as function of the beam energy for U^{92+} → N$_2$ collisions [28]. The strong variation observed is a result of the energy dependence of the different population mechanism

Table 4. Result for the Lyman-α_1 transition energy in U^{91+} (compare text) [19]. All values are given in eV

	ΔE (β)	ΔE (FIT)	ΔE (geometry)	Lyman-α_1
49 MeV/u	± 3	± 13.2	± 8	102 162 ± 15.7
68 MeV/u	± 2.3	± 14.3	± 9	102 179 ± 17.0
Result	± 2.6	± 9.7	± 8.5	102 170.7 ± 13.2

In Table 4, the final experimental result for the Lyman-α_1 transition energy as deduced from the fitting procedure is given. Only the *Lyman* α_1 centroid energy allows a direct comparison with the groundstate *Lamb-shift* prediction since here only one transition, i.e. the decay of the $2p_{3/2}$ state, contributes to the observed line. Also, due to the intrinsic resolution of the detectors used (between 500 eV and 650 eV at 122 keV) the **M1** line blend of the Lyman-α_2 decay cannot be resolved experimentally since the $2s_{1/2} - 2p_{1/2}$ line spacing amounts to 70 eV (classical Lamb shift). In addition the energy dependence of the population processes for the j=1/2 of the L-shell leads to a considerable shift of the Lyman-α_2-**M1** centroid as function of energy (see Fig. 14). From the result for the Lyman-α_1 transition energy quoted in Table 4 the final experimental Lamb shift value follows from comparison with the Dirac eigenvalue for the 1s state of -132279.96 eV (see Table 1).

3.4 Experimental Results in Comparison with Theory

In Fig. 15, the experimental results for the ground-state Lamb shift in H-like ions are given and compared with the theoretical predictions [3] (full line). For

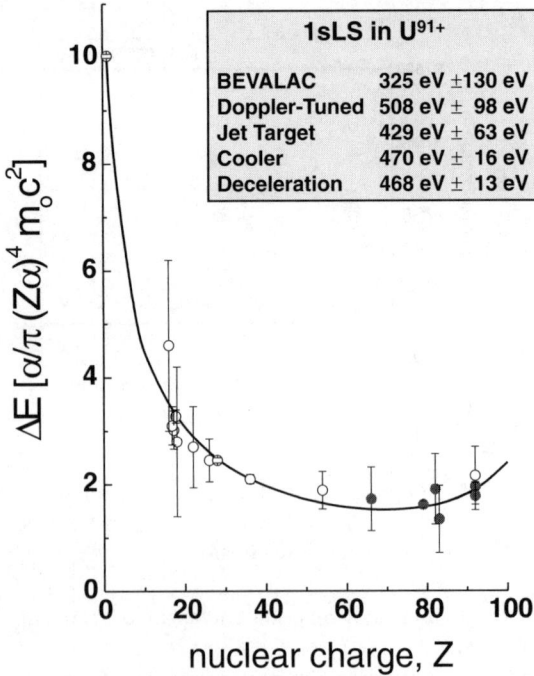

Fig. 15. All available experimental results for the 1s Lamb shift in high-Z ions in comparison with theoretical predictions [3]. In the inset the available Lamb shift data for U^{91+} are shown [13,17,19,32,33]

comparison, the data shown in the figure are given in units of the function $F(Z\alpha)$ (Eq. 1). The solid symbols depict the results from the SIS/ESR facility. Over the whole range of nuclear charges an excellent overall agreement between experiment and theory is observed. For the regime of the high-Z ions ($Z > 54$) most of the results provide a test of the ground-state Lamb shift contribution at the level of 30%. Only the results from the gasjet target (for uranium) and from the electron cooler (for gold and for uranium) have a considerably higher accuracy. Up to now most of the Lamb shift experiments for high-Z ions were performed for hydrogen-like uranium. Therefore, these data are given in addition separately in the inset of Fig. 15 in comparison with the theoretical prediction (the data collected at the ESR are represented by solid symbols). The figure demonstrates the substantial improvement by almost one order of magnitude achieved at the storage ring as compared to earlier experiments conducted at the BEVALAC accelerator [32,33].

The theoretical 1s Lamb shift of 463.95(50) eV [10] has to be compared with the experimental value from the jet-target of 468 eV $\pm$ 13 eV [19]. The theoretical value was calculated including all second order (in α) contributions which until recently were the largest sources of the theoretical uncertainty [34–36]. However, an experimental test of these contributions still requires a further improvement of the experimental accuracy by almost one order of magnitude in order to approach an absolute precision of 1eV. Note, that the finite nuclear size effect contributes more than 40% to the total Lamb shift correction. Although the uncertainties introduced by the latter effect (0.3 eV [8]) are much smaller than the present experimental accuracy, they may prevent, for the particular case of uranium, a direct test of QED with a precision of 1 eV or better. For this purpose, an experiment using ^{208}Pb appears to be most appropriate because the extended nuclear size of this double magic nucleus is much better known.

3.5 Towards an Accuracy of 1 eV

In the most recent 1s Lamb shift experiment for U^{91+} a ground state Lamb shift of 468 ± 13 eV was obtained which is the most precise test of quantum electrodynamics for a single electron system in the strong field regime and is at the threshold of a meaningful test of higher-order QED contributions. By using decelerated beams, further progress towards an absolute accuracy of 1 eV may be anticipated. In order to obtain a significantly improved precision, future experiments will also use a highly redundant setup at the jet-target and will focus in addition on decelerated ions combined with the crystal spectrometers [37,38] or bolometers [39,40] which presently are under construction.

As an example, a new kind of x-ray spectrometer, set up in the **FO**cussing **C**ompensated **A**symmetric **L**aue (**FOCAL**) geometry, has been developed for this purpose [38] (see Fig. 16). The spectrometer serves in measuring small wavelength differences between the fast moving x-ray source, represented by the circulating ions in the ESR, and a stationary calibration source. It is designed for energies between 50 and 10 keV or wavelengths between 25 and 12 pm leading to Bragg angles of less than 4° for a Si(220) crystal. In an future experiment, this focusing transmission crystal spectrometer will be combined with an segmented germanium x-ray detectors. Such an position-sensitive detector permits the measurement of an position spectrum at the focus of the spectrometer which is wide enough to investigate the interesting energy regime simultaneously. In

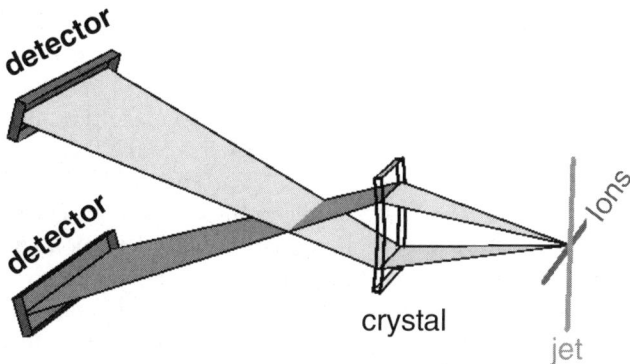

Fig. 16. Experimental arrangement of the FOCAL spectrometer at the ESR jet target [38]

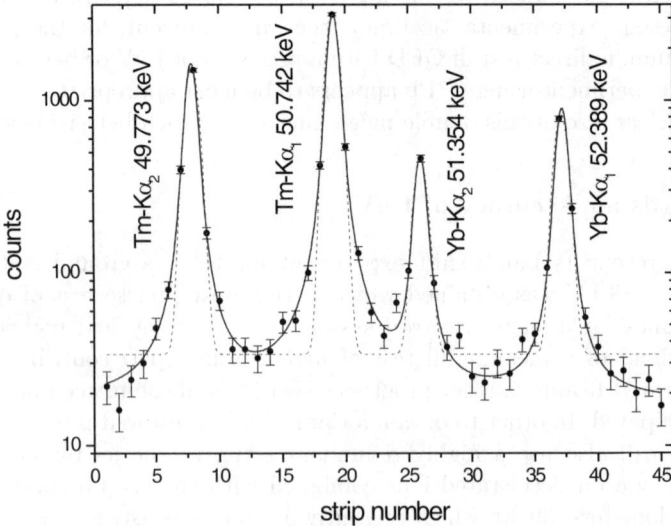

Fig. 17. Result obtained with the germanium microstrip detector mounted at the FOCAL spectrometer. The intensity pattern as a function of the position (energy) identifies well resolved the two components of the Kα-doublet of Tm as well as those of Yb [41]. The solid line refers to a least square fit by using Voigt profiles and the dashed line refers to Gaussian distributions

addition, the good energy resolution of such a detector enables discrimination against background events of the recorded spectra arising from various sources.

Very recently such a microstrip detector system, developed at the Forschungs-zentrum Jülich [41], with a position resolution of close to 200 μm has become available and been tested in combination with the FOCAL spectrometer using an intense radioactive ^{169}Yb source. Even without any strict conditions on the photon energies for the individual strips, the intensity pattern observed with the microstrip detector as function of the position (i.e. strip number) identifies clearly the two x-ray lines of the Kα-doublet from Tm and Yb (Fig. 17) which are separated by approximately 970 eV and 1030 eV, respectively. This demonstrates that in combination with the FOCAL spectrometer [38], an energy resolution better than 100 eV can be achieved along with a high detection efficiency.

Finally we like to emphasis that the spectral line shapes can not be described with Gaussian function as it is obvious from the Fig. 17. A Voigt profile (convolution of Lorentzian and Gaussian) should be used instead. This means that the achieved resolution of the FOCAL spectrometer in combination with the microstrip germanium detector allows us to be sensitive to the natural linewidths of the transitions.

4   Summary and Outlook

In summary, the cooled heavy-ion beams of the ESR storage ring offer excellent experimental conditions for a precise study of the effects of QED in the ground-state of high-Z H-like ions. This has been demonstrated within the first series of experiments conducted at the gasjet target as well as at the electron cooler device. For the case of the 1s-Lamb shift in hydrogen-like uranium, the achieved accuracy of ±13 eV is already a substantial improvement by almost one order of magnitude as compared to earlier experiments conducted at the BEVALAC accelerator [32,33], and the available results from the ESR are now at the threshold of a real test of higher-order QED contributions. By using decelerated beams, further progress towards an absolute accuracy of 1 eV may be anticipated. For this purpose an improvement of the spectral resolution is needed in addition.

As has been demonstrated in the first experiment with decelerated ions, the deceleration mode not only reduces the uncertainties in the Doppler corrections but provides in particular a very efficient production of characteristic projectile radiation. Along with a strongly enhanced injection efficiency into the ESR this will allow one to implement at the gasjet target high-resolution x-ray detection devices such as crystal spectrometers [37] or bolometers [40] which presently are under construction. Such high-resolution detectors are finally required in order to reach a precision close to 1 eV.

Acknowledgment

The work at the ESR was done in teamwork with F. Bosch, R.W. Dunford, P. Indelicato, A. Krämer, D. Liesen, A. Orsic Muthig, C. Kozhuharov, T. Ludziejewski, P.H. Mokler, U. Spillmann, Z. Stachura, S. Tachenov, A. Warczak, and with the members of the ESR team, K. Beckert, P. Beller, B. Franzke, F Nolden, and M. Steck. The close collaboration with D. Protic and Th. Krings is gratefully acknowledged. The authors would like to thank T. Beier, H.J. Kluge, and V.M. Shabaev for stimulating discussions.

References

1. W.E. Lamb and R.C. Retherford, Phys. Rev. 72, 241(1947).
2. P.J. Mohr, At. Data and Nucl. Data Tables **29**, 453 (1983).
3. W.R. Johnson and G. Soff, At. Data and Nucl. Data Tables **33**, 405 (1985).
4. S. Bourzeix, B. de Beauvoir, F. Nez, M.D. Plimmer, F. de Tomasi, L. Julien, F. Biraden, D.N. Stacey, Phys. Rev. Lett. **76**, 384 (1996).
5. B. de Beauvoir et al., Phys. Rev. Lett. **78**, 440 (1997).
6. Th. Udem et al., Phys. Rev. Lett. **79**, 2646 (1997).
7. P.J. Mohr, Phys. Rev. A **46**, 4421 (1993).
8. H. Persson , S. Salomonson, P. Sunnergren, I. Lindgren, M.G.H. Gustavsson, Hyperfine Interaction **108**, 3 (1997).
9. H. Persson, I. Lindgren, L. Labzowsky, G. Plunien, T. Beier, G. Soff, Phys. Rev. A**54**, 2805 (1996).

10. V.A. Yerokhin and V.M. Shabaev, Phys. Rev. A**64**, 062507 (2001).
11. T. Beier, P.J. Mohr, H. Persson, G. Plunien, M. Greiner, G. Soff, Phys. Lett. A**236**, 329 (1997).
12. P. Indelicato, private communication (1998).
13. Th. Stöhlker, P. H. Mokler, K. Beckert, F. Bosch, H. Eickhoff, B. Franzke, M. Jung, T. Kandler, O. Klepper, C. Kozhuharov, R. Moshammer, F. Nolden, H. Reich, P. Rymuza, P. Spädtke, M. Steck, Phys. Rev. Lett. **71**, 2184 (1993).
14. P. H. Mokler, Th. Stöhlker, C. Kozhuharov, R. Moshammer, P. Rymuza, F. Bosch, T. Kandler, Physica Scripta Vol. T**51**, 28 (1994).
15. H. F. Beyer, D. Liesen, F. Bosch, K. D. Finlayson, M. Jung, O. Klepper, R. Moshammer, K. Beckert, H. Eickhoff, B. Franzke, F. Nolden, P. Spädtke, M. Steck, G. Menzel, R. D. Deslattes, Phys. Lett. A **184**, 435 (1994).
16. H.F. Beyer, IEEE Trans. Instr. Meas. **44**, 510 (1995).
17. H.F. Beyer, G. Menzel, D. Liesen, A. Gallus, F. Bosch, R. Deslattes, P. Indelicato, Th. Stöhlker, O. Klepper, R. Moshammer, F. Nolden, H. Eickhoff, B. Franzke, M. Steck, Z. Phys. D **35**, 169 (1995).
18. Th. Stöhlker, Physica Scripta T**73**, 29 (1997).
19. Th. Stöhlker, P.H. Mokler, F. Bosch, R.W. Dunford, O. Klepper, C. Kozhuharov, T. Ludziejewski, B. Franzke, F. Nolden, H. Reich, P. Rymuza, Z. Stachura, M. Steck, P. Swiat, and A. Warczak, Phys. Rev. Lett. **85**, 3109 (2000).
20. D. Liesen, H.F. Beyer, and G. Menzel, Comments At. Mol. Phys. **32**, 23 (1995).
21. H. F. Beyer, H.-J. Kluge, V.P. Shevelko, *X-ray Radiation of Highly Charged Ions* (Springer, Berlin, Heidelberg) (1996).
22. P. H. Mokler and Th. Stöhlker, Physics of Highly-Charged Heavy Ions Revealed by Storage/Cooler Rings, Advances in Atomic, Molecular and Optical **37**, 297 (1996).
23. M. Steck, K. Beckert, F. Bosch, H. Eickhoff, B. Franzke, O. Klepper, R. Moshammer, F. Nolden, P. Spädtke, and Th. Winkler, *Proc 4th Europ. Part. Accel. Conf., London 1994*, ed. V. Suller and Ch. Petit-Jean-Genaz (World Scientific, Singapore 1994), p. 1197.
24. B. Franzke, *Information about ESR Paramteres*, GSI-ESR/TN-86-01, 1986 (Internal Report).
25. A. Gumberidze et al., to be published (2002).
26. P.H. Mokler, Th. Stöhlker, R.W. Dunford, A. Gallus, T. Kandler, G. Menzel, T. Prinz, P. Rymuza, Z. Stachura, P. Swiat, A.Warczak, Z. Phys. D **35**, 274(1995)
27. Th. Stöhlker, T.Ludziejewski, H. Reich, F. Bosch, R.W. Dunford, J. Eichler, B. Franzke, C. Kozhuharov, G. Menzel, P.H. Mokler, F. Nolden, P. Rymuza, Z. Stachura, M. Steck, P. Swiat, A. Warczak, Phys. Rev. A **58**, 2043 (1998).
28. Th. Stöhlker, Habilitation Thesis, University Frankfurt, unpublished (1998).
29. L.C. Longoria et al., Nucl. Instr. Meth. A **299**, 208 (1990).
30. P. Swiat, A. Warczak, F. Bosch, B. Franzke, O. Klepper, C. Kozhuharov, G. Menzel, P.H. Mokler, H. Reich, Th. Stöhlker, R.W. Dunford, P. Rymuza, T. Ludziejewski, Z. Stachura, Phys. Scrip. T**80**, 326 (1999).
31. M. Steck, private communication.
32. J. P. Briand, P. Chevallier, P.Indelicato, K.P. Ziock, and D. Dietrich, Phys. Rev. Lett. **65**, 2761 (1990).
33. J. H. Lupton, D.D. Dietrich, C.J. Hailey, R.E. Stewart, K.P. Ziock, Phys. Rev. A **50**, 2150 (1994).
34. A. Mitrushenkov, L. Labzowsky, I. Lindgren, H. Persson and S. Salomonoson, Phys. Lett. A **200**, 51 (1995).

35. S. Mallampalli and J. Sapirstein, Phys. Rev. Lett. **80**, 5297 (1998)
36. V. A. Yerokhin, Phys. Rev. A**62**, 012508 (2000).
37. H. F. Beyer, *Physics with Multiply Charged Ions*, Edited by D. Liesen, (Plenum Press, New York 1995).
38. H. Beyer, Nucl. Instr. Meth A **400**, 137 (1997).
39. P. Egelhof, H. F. Beyer, D. McCammon, F. v. Feilitzsch, A. v. Kienlin, H.-J. Kluge, D. Liesen, J. Meier, S. H. Moseley, Th. Stöhlker, Nucl. Instr. Meth. A**370**, 26(1996).
40. P. Egelhof, Adv. in Solid State Phys. **39**, 61 (1999).
41. D. Protic, Th. Stöhlker, H. F. Beyer, J. Bojowald, G. Borchert, A. Hamacher, C. Kozhuharov, T. Ludziejewski, X. Ma, I. Mohos, IEEE Trans. Nucl. Sci. **48**, 1048 (2001).

Part IV

Testing Quantum Electrodynamics

Simple Atoms, Quantum Electrodynamics, and Fundamental Constants

Savely G. Karshenboim

Max-Planck-Institut für Quantenoptik, 85748 Garching, Germany
D.I. Mendeleev Institute for Metrology (VNIIM), St. Petersburg 198005, Russia

Abstract. This review is devoted to precision physics of simple atoms. The atoms can essentially be described in the framework of quantum electrodynamics (QED), however, the energy levels are also affected by the effects of the strong interaction due to the nuclear structure. We pay special attention to QED tests based on studies of simple atoms and consider the influence of nuclear structure on energy levels. Each calculation requires some values of relevant fundamental constants. We discuss the accurate determination of the constants such as the Rydberg constant, the fine structure constant and masses of electron, proton and muon etc.

1 Introduction

Simple atoms offer an opportunity for high accuracy calculations within the framework of quantum electrodynamics (QED) of bound states. Such atoms also possess a simple spectrum and some of their transitions can be measured with high precision. Twenty, thirty years ago most of the values which are of interest for the comparison of theory and experiment were known experimentally with a higher accuracy than from theoretical calculations. After a significant theoretical progress in the development of bound state QED, the situation has reversed. A review of the theory of light hydrogen-like atoms can be found in [1], while recent advances in experiment and theory have been summarized in the Proceedings of the International Conference on Precision Physics of Simple Atomic Systems (2000) [2].

Presently, most limitations for a comparison come directly or indirectly from the experiment. Examples of a direct experimental limitation are the $1s - 2s$ transition and the $1s$ hyperfine structure in positronium, whose values are known theoretically better than experimentally. An indirect *experimental* limitation is a limitation of the precision of a *theoretical* calculation when the uncertainty of such calculation is due to the inaccuracy of fundamental constants (e.g. of the muon-to-electron mass ratio needed to calculate the $1s$ hyperfine interval in muonium) or of the effects of strong interactions (like e.g. the proton structure for the Lamb shift and $1s$ hyperfine splitting in the hydrogen atom). The knowledge of fundamental constants and hadronic effects is limited by the experiment and that provides experimental limitations on theory.

This is not our first brief review on simple atoms (see e.g. [3,4]) and to avoid any essential overlap with previous papers, we mainly consider here the most recent progress in the precision physics of hydrogen-like atoms since the publication of the Proceedings [2]. In particular, we discuss

- Lamb shift in the hydrogen atom;
- hyperfine structure in hydrogen, deuterium and helium ion;
- hyperfine structure in muonium and positronium;
- g factor of a bound electron.

We consider problems related to the accuracy of QED calculations, hadronic effects and fundamental constants.

   These atomic properties are of particular interest because of their applications beyond atomic physics. Understanding of the Lamb shift in hydrogen is important for an accurate determination of the Rydberg constant Ry and the proton charge radius. The hyperfine structure in hydrogen, helium-ion and positronium allows, under some conditions, to perform an accurate test of bound state QED and in particular to study some higher-order corrections which are also important for calculating the muonium hyperfine interval. The latter is a source for the determination of the fine structure constant α and muon-to-electron mass ratio. The study of the g factor of a bound electron lead to the most accurate determination of the proton-to-electron mass ratio, which is also of interest because of a highly accurate determination of the fine structure constant.

2   Rydberg Constant and Lamb Shift in Hydrogen

About fifty years ago it was discovered that in contrast to the spectrum predicted by the Dirac equation, there are some effects in hydrogen atom which split the $2s_{1/2}$ and $2p_{1/2}$ levels. Their splitting known as the Lamb shift (see Fig. 1) was successfully explained by quantum electrodynamics. The QED effects lead to a tiny shift of energy levels and for thirty years this shift was studied by means of

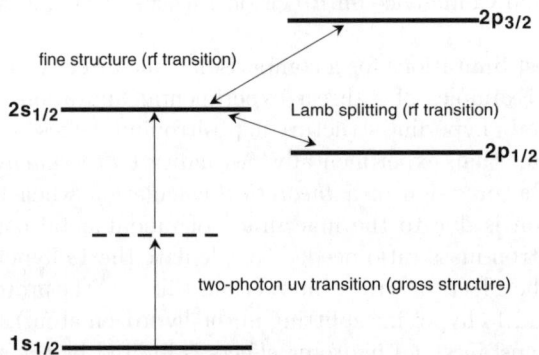

Fig. 1. Spectrum of the hydrogen atom (not to scale). The hyperfine structure is neglected. The label *rf* stands for radiofrequency intervals, while *uv* is for ultraviolet transitions

microwave spectroscopy (see e.g. [5,6]) measuring either directly the splitting of the $2s_{1/2}$ and $2p_{1/2}$ levels or a bigger splitting of the $2p_{3/2}$ and $2s_{1/2}$ levels (fine structure) where the QED effects are responsible for approximately 10% of the fine-structure interval.

The recent success of two-photon Doppler-free spectroscopy [7] opens another way to study QED effects directed by high-resolution spectroscopy of gross-structure transitions. Such a transition between energy levels with different values of the principal quantum number n is determined by the Coulomb-Schrödinger formula

$$E(nl) = -\frac{(Z\alpha)^2 mc^2}{2n^2} , \tag{1}$$

where Z is the nuclear charge in units of the proton charge, m is the electron mass, c is the speed of light, and α is the fine structure constant. For any interpretation in terms of QED effects one has to determine a value of the Rydberg constant

$$Ry = \frac{\alpha^2 mc}{2h} , \tag{2}$$

where h is the Planck constant. Another problem in the interpretation of optical measurements of the hydrogen spectrum is the existence of a few levels which are significantly affected by the QED effects. In contrast to radiofrequency measurements, where the $2s - 2p$ splitting was studied, optical measurements have been performed with several transitions involving $1s$, $2s$, $3s$ etc. It has to be noted that the theory of the Lamb shift for levels with $l \neq 0$ is relatively simple, while theoretical calculations for s states lead to several serious complications. The problem of the involvement of few s levels has been solved by introducing an auxiliary difference [8]

$$\Delta(n) = E_L(1s) - n^3 E_L(ns) , \tag{3}$$

for which theory is significantly simpler and more clear than for each of the s states separately.

Combining theoretical results for the difference [9] with measured frequencies of two or more transitions one can extract a value of the Rydberg constant and of the Lamb shift in the hydrogen atom. The most recent progress in determination of the Rydberg constant is presented in Fig. 2 (see [7,10] for references).

Presently the optical determination [7,4] of the Lamb shift in the hydrogen atom dominates over the microwave measurements [5,6]. The extracted value of the Lamb shift has an uncertainty of 3 ppm. That ought to be compared with the uncertainty of QED calculations (2 ppm) [11] and the uncertainty of the contributions of the nuclear effects. The latter has a simple form

$$\Delta E_{\text{charge radius}}(nl) = \frac{2(Z\alpha)^4 mc^2}{3n^3} \left(\frac{mcR_p}{\hbar}\right)^2 \delta_{l0} . \tag{4}$$

To calculate this correction one has to know the proton rms charge radius R_p with sufficient accuracy. Unfortunately, it is not known well enough [11,3] and

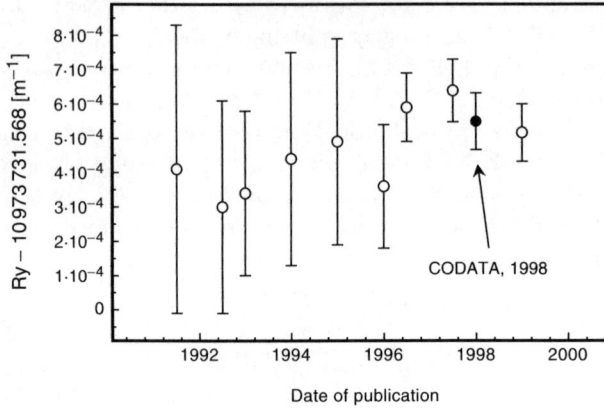

Fig. 2. Progress in the determination of the Rydberg constant by two-photon Doppler-free spectroscopy of hydrogen and deuterium. The label *CODATA, 1998* stands for the recommended value of the Rydberg constant ($Ry = 10\,973\,731.568\,549(83)$ m^{-1} [10])

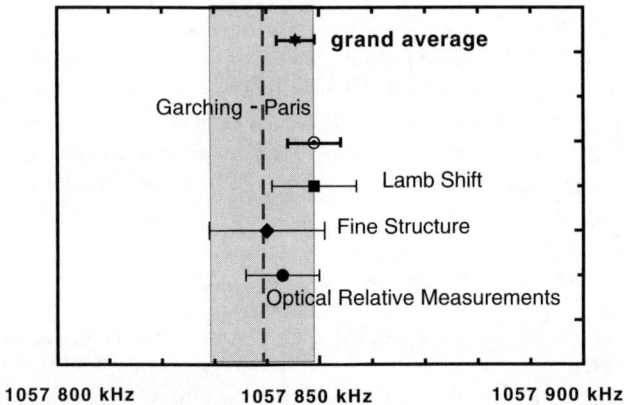

Fig. 3. Measurement of the Lamb shift in hydrogen atom. Theory is presented according to [11]. The most accurate value comes from comparison of the $1s - 2s$ transition at MPQ (Garching) and the $2s - ns/d$ at LKB (Paris), where $n = 8, 10, 12$. Three results are shown: for the average values extracted from direct Lamb shift measurements, measurements of the fine structure and a comparison of two optical transitions within a single experiment. The filled part is for the theory

leads to an uncertainty of 10 ppm for the calculation of the Lamb shift. It is likely that a result for R_p from the electron-proton elastic scattering [12] cannot be improved much, but it seems to be possible to significantly improve the accuracy of the determination of the proton charge radius from the Lamb-shift experiment on muonic hydrogen, which is now in progress at PSI [13].

Table 1. Hyperfine structure in light hydrogen-like atoms: QED and nuclear contributions ΔE(Nucl). The numerical results are presented for the frequency E/h

Atom, state	E_{HFS}(exp) [kHz]	Ref.	E_{HFS}(QED) [kHz]	ΔE(Nucl) [ppm]
Hydrogen, $1s$	1 420 405.751 768(1)	[16,17]	1 420 452	- 33
Deuterium, $1s$	327 384.352 522(2)	[18]	327 339	138
Tritium, $1s$	1 516 701.470 773(8)	[19]	1 516 760	- 36
^{3}He$^+$ ion, $1s$	- 8 665 649.867(10)	[20]	- 8 667 494	- 213
Hydrogen, $2s$	177 556.860(15)	[21,22]	177 562.7	-32
Hydrogen, $2s$	177 556.785(29)	[23]		- 33
Hydrogen, $2s$	177 556.860(50)	[24]		- 32
Deuterium, $2s$	40 924.439(20)	[25]	40 918.81	137
^{3}He$^+$ ion, $2s$	- 1083 354.980 7(88)	[26]	- 1083 585.3	- 213
^{3}He$^+$ ion, $2s$	- 1083 354.99(20)	[27]		- 213

Table 2. Comparison of bound QED and nuclear corrections to the $1s$ hyperfine interval. The QED term ΔE(QED) contains only bound-state corrections and the contribution of the anomalous magnetic moment of electron is excluded. The nuclear contribution ΔE(Nucl) has been found via comparison of experimental results with pure QED values (see Table 1)

Atom	ΔE(QED) [ppm]	ΔE(Nucl) [ppm]
Hydrogen	23	- 33
Deuterium	23	138
Tritium	23	- 36
^{3}He$^+$ ion	108	- 213

3   Hyperfine Structure and Nuclear Effects

A similar problem of interference of nuclear structure and QED effects exists for the $1s$ and $2s$ hyperfine structure in hydrogen, deuterium, tritium and helium-3 ion. The magnitude of nuclear effects entering theoretical calculations is at the level from 30 to 200 ppm (depending on the atom) and their understanding is unfortunately very poor [11,14,15]. We summarize the data in Tables 1 and 2 (see [15][1] for detail).

The leading term (so-called Fermi energy E_F) is a result of the nonrelativistic interaction of the Dirac magnetic moment of electron with the actual nuclear magnetic moment. The leading QED contribution is related to the anomalous magnetic moment and simply rescales the result ($E_F \to E_F \cdot (1 + a_e)$). The result

---

[1] A misprint in a value of the nuclear magnetic moment of helium-3 (it should be $\mu/\mu_B = -1.158\,740\,5$ instead of $\mu/\mu_B = -1.158\,750\,5$) has been corrected and some results on helium received minor shifts which are essentially below uncertainties

of the QED calculations presented in Table 1 is of the form

$$E_{\text{HFS}}(\text{QED}) = E_F \cdot (1 + a_e) + \Delta E(\text{QED}) , \qquad (5)$$

where the last term which arises from bound-state QED effects for the $1s$ state is given by

$$
\Delta E_{1s}(\text{QED}) = E_F \times \left\{ \frac{3}{2}(Z\alpha)^2 + \alpha(Z\alpha)\left(\ln 2 - \frac{5}{2}\right) \right.
$$
$$
+ \frac{\alpha(Z\alpha)^2}{\pi}\left[-\frac{2}{3}\ln\frac{1}{(Z\alpha)^2}\left(\ln\frac{1}{(Z\alpha)^2}\right.\right.
$$
$$
\left. + 4\ln 2 - \frac{281}{240}\right) + 17.122\,339\ldots
$$
$$
\left.\left. - \frac{8}{15}\ln 2 + \frac{34}{225}\right] + 0.7718(4)\,\frac{\alpha^2(Z\alpha)}{\pi} \right\} . \qquad (6)
$$

This term is in fact smaller than the nuclear corrections as it is shown in Table 2 (see [15] for detail). A result for the $2s$ state is of the same form with slightly different coeffitients [15].

From Table 1 one can learn that in relative units the effects of nuclear structure are about the same for the $1s$ and $2s$ intervals (33 ppm for hydrogen, 138 ppm for deuterium and 213 ppm for helium-3 ion). A reason for that is the factorized form of the nuclear contributions in leading approximation (cf. (4))

$$\Delta E(\text{Nucl}) = A(\text{Nucl}) \times |\Psi_{nl}(\mathbf{r} = 0)|^2 \qquad (7)$$

i.e. a product of the nuclear-structure parameter $A(\text{Nucl})$ and a the wave function at the origin

$$|\Psi_{nl}(\mathbf{r} = 0)|^2 = \frac{1}{\pi}\left(\frac{(Z\alpha)m_R c}{n\hbar}\right)^3 \delta_{l0} , \qquad (8)$$

which is a result of a pure atomic problem (a nonrelativistic electron bound by the Coulomb field). The nuclear parameter $A(\text{Nucl})$ depends on the nucleus (proton, deutron $etc.$) and effect (hyperfine structure, Lamb shift) under study, but does not depend on the atomic state.

Two parameters can be changed in the wave function:

- the principle quantum number $n = 1, 2$ for the $1s$ and $2s$ states;
- the reduced mass of a bound particle for conventional (electronic) atoms ($m_R \simeq m_e$) and muonic atoms ($m_R \simeq m_\mu$).

The latter option was mentioned when considering determination of the proton charge radius via the measurement of the Lamb shift in muonic hydrogen [13]. In the next section we consider the former option, comparison of the $1s$ and $2s$ hyperfine interval in hydrogen, deuterium and ion $^3\text{He}^+$.

Table 3. Theory of the specific difference $D_{21} = 8E_{\text{HFS}}(2s) - E_{\text{HFS}}(1s)$ in light hydrogen-like atoms (see [15] for detail). The numerical results are presented for the frequency D_{21}/h

Contribution	Hydrogen [kHz]	Deuterium [kHz]	^{3}He$^+$ ion [kHz]
D_{21}(QED3)	48.937	11.305 6	-1 189.252
D_{21}(QED4)	0.018(3)	0.004 3(5)	-1.137(53)
D_{21}(nucl)	-0.002	0.002 6(2)	0.317(36)
D_{21}(theo)	48.953(3)	11.312 5(5)	-1 190.072(63)

4   Hyperfine Structure of the 2s State in Hydrogen, Deuterium, and Helium-3 Ion

Our consideration of the $2s$ hyperfine interval is based on a study of the specific difference

$$D_{21} = 8 \cdot E_{\text{HFS}}(2s) - E_{\text{HFS}}(1s) , \qquad (9)$$

where any contribution which has a form of (7) should vanish.

The difference (9) has been studied theoretically in several papers long ago [28–30]. A recent study [31] shown that some higher-order QED and nuclear corrections have to be taken into account for a proper comparison of theory and experiment. The theory has been substantially improved [15,32] and it is summarized in Table 3. The new issues here are most of the fourth-order QED contributions (D_{21}(QED4)) of the order $\alpha(Z\alpha)^3$, $\alpha^2(Z\alpha)^4$, $\alpha(Z\alpha)^2m/M$ and $(Z\alpha)^3m/M$ (all are in units of the $1s$ hyperfine interval) and nuclear corrections (D_{21}(nucl)). The QED corrections up to the third order (D_{21}(QED3)) and the fourth-order contribution of the order $(Z\alpha)^4$ have been known for a while [28–30,33].

For all the atoms in Table 3 the hyperfine splitting in the ground state was measured more accurately than for the $2s$ state. All experimental results but one were obtained by direct measurements of microwave transitions for the $1s$ and $2s$ hyperfine intervals. However, the most recent result for the hydrogen atom has been obtained by means of laser spectroscopy and measured transitions lie in the ultraviolet range [21,22]. The hydrogen level scheme is depicted in Fig. 4. The measured transitions were the singlet-singlet ($F = 0$) and triplet-triplet ($F = 1$) two-photon $1s - 2s$ ultraviolet transitions. The eventual uncertainty of the hyperfine structure is to 6 parts in 10^{15} of the measured $1s - 2s$ interval. The optical result in Table 1 is a preliminary one and the data analysis is still in progress.

The comparison of theory and experiment for hydrogen and helium-3 ion is summarized in Figs. 5 and 6.

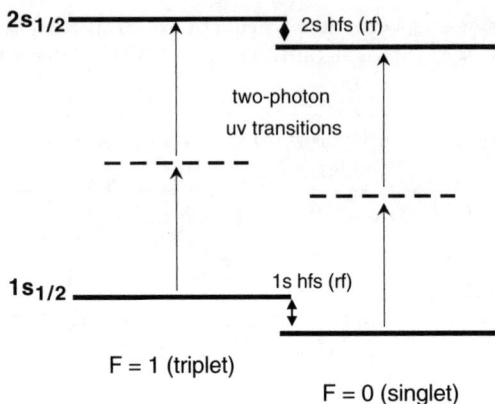

Fig. 4. Level scheme for an optical measurement of the hyperfine structure (*hfs*) in the hydrogen atom (not to scale) [22]. The label *rf* stands here for radiofrequency intervals, while *uv* is for ultraviolet transitions

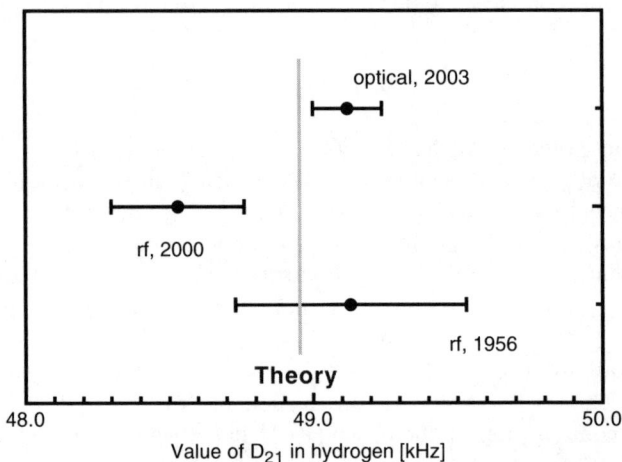

Fig. 5. Present status of measurements of D_{21} in the hydrogen atom. The results are labeled with the date of the measurement of the 2s hyperfine structure. See Table 1 for references

5   Hyperfine Structure in Muonium and Positronium

Another possibility to eliminate nuclear structure effects is based on studies of nucleon-free atoms. Such an atomic system is to be formed of two leptons. Two atoms of the sort have been produced and studied for a while with high accuracy, namely, muonium and positronium.

- Muonium is a bound system of a positive muon and electron. It can be produced with the help of accelerators. The muon lifetime is $2.2 \cdot 10^{-6}$ sec. The most accurately measured transition is the 1s hyperfine structure. The

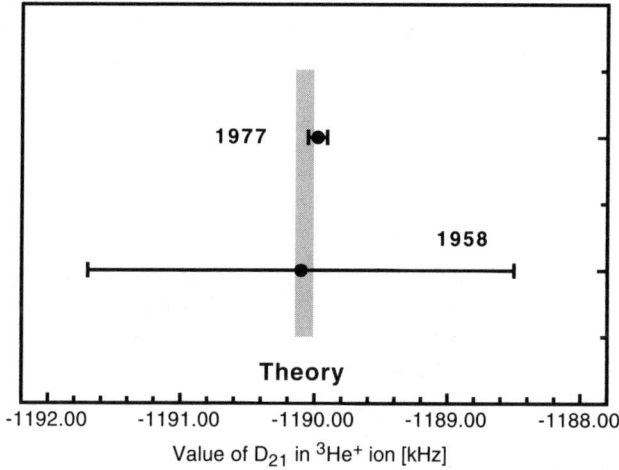

Fig. 6. Present status of measurements of D_{21} in the helium ion $^3\mathrm{He}^+$. See Table 1 for references

two-photon $1s - 2s$ transition was also under study. A detailed review of muonium physics can be found in [34].

- Positronium can be produced at accelerators or using radioactive positron sources. The lifetime of positronium depends on its state. The lifetime for the $1s$ state of parapositronium (it annihilates mainly into two photons) is $1.25 \cdot 10^{-10}$ sec, while orthopositronium in the $1s$ state has a lifetime of $1.4 \cdot 10^{-7}$ s because of three-photon decays. A list of accurately measured quantities contains the $1s$ hyperfine splitting, the $1s - 2s$ interval, $2s - 2p$ fine structure intervals for the triplet $1s$ state and each of the four $2p$ states, the lifetime of the $1s$ state of para- and orthopositronium and several branchings of their decays. A detailed review of positronium physics can be found in [35].

Here we discuss only the hyperfine structure of the ground state in muonium and positronium. The theoretical status is presented in Tables 4 and 5. The theoretical uncertainty for the hyperfine interval in positronium is determined only by the inaccuracy of the estimation of the higher-order QED effects. The uncertainty budget in the case of muonium is more complicated. The biggest source is the calculation of the Fermi energy, the accuracy of which is limited by the knowledge of the muon magnetic moment or muon mass. It is essentially the same because the g factor of the free muon is known well enough [45]. The uncertainty related to QED is determined by the fourth-order corrections for muonium ($\Delta E(QED4)$) and the third-order corrections for positronium ($\Delta E(QED3)$). These corrections are related to essentially the same diagrams (as well as the $D_{21}(QED4)$ contribution in the previous section). The muonium uncertainty is due to the calculation of the recoil corrections of the order of $\alpha(Z\alpha)^2 m/M$ [42,46] and $(Z\alpha)^3 m/M$, which are related to the third-order contributions [42] for positronium since $m = M$.

Table 4. Theory of the 1s hyperfine splitting in muonium. The numerical results are presented for the frequency E/h. The calculations [36] have been performed for $\alpha^{-1} = 137.035\,999\,58(52)$ [37] and $\mu_\mu/\mu_p = 3.183\,345\,17(36)$ which was obtained from the analysis of the data on Breit-Rabi levels in muonium [38,39] (see Sect. 6) and precession of the free muon [40]. The numerical results are presented for the frequency E/h

Term	Fractional contribution	ΔE [kHz]
E_F	1.000 000 000	4.459 031.83(50)(3)
a_e	0.001 159 652	5 170.926(1)
QED2	- 0.000 195 815	- 873.147
QED3	- 0.000 005 923	- 26.410
QED4	- 0.000 000 123(49)	- 0.551(218)
Hadronic	0.000 000 054(1)	0.240(4)
Weak	- 0.000 000 015	- 0.065
Total	1.000 957 830(49)	4 463 302.68(51)(3)(22)

Table 5. Theory of the 1s hyperfine interval in positronium. The numerical results are presented for the frequency E/h. The calculation of the second order terms was completed in [41], the leading logarithmic contributions were found in [42], while next-to-leading logarithmic terms in [43]. The uncertainty is presented following [44]

Term	Fractional contribution	ΔE [MHz]
E_F	1.000 000 0	204 386.6
QED1	- 0.004 919 6	-1 005.5
QED2	0.000 057 7	11.8
QED3	- 0.000 006 1(22)	- 1.2(5)
Total	0.995 132 1(22)	203 391.7(5)

The muonium calculation is not completely free of hadronic contributions. They are discussed in detail in [36,47,48] and their calculation is summarized in Fig. 7. They are small enough but their understanding is very important because of the intensive muon sources expected in future [49] which might allow to increase dramatically the accuracy of muonium experiments.

A comparison of theory versus experiment for muonium is presented in the summary of this paper. Present experimental data for positronium together with the theoretical result are depicted in Fig. 8.

6 g Factor of Bound Electron and Muon in Muonium

Not only the spectrum of simple atoms can be studied with high accuracy. Other quantities are accessible to high precision measurements as well among them the

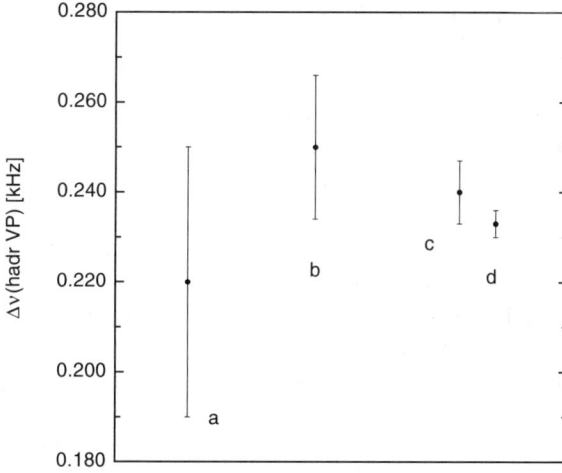

Fig. 7. Hadronic contributions to HFS in muonium. The results are taken: a from [50], b from [51], c from [52] and d from [36,47]

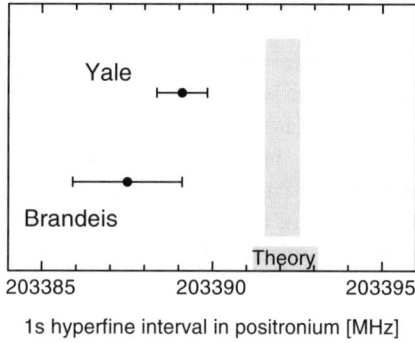

Fig. 8. Positronium hyperfine structure. The Yale experiment was performed in 1984 [53] and the Brandeis one in 1975 [54]

atomic magnetic moment. The interaction of an atom with a weak homogeneous magnetic field can be expressed in terms of an effective Hamiltonian. For muonium such a Hamiltonian has the form

$$\mathcal{H} = \frac{e\hbar}{2m_e}g'_e\left(\mathbf{s}_e \cdot \mathbf{B}\right) - \frac{e\hbar}{2m_N}g'_\mu\left(\mathbf{s}_\mu \cdot \mathbf{B}\right) + \Delta E_{\mathrm{HFS}}\left(\mathbf{s}_e \cdot \mathbf{s}_\mu\right), \qquad (10)$$

where $\mathbf{s}_{e(\mu)}$ stands for spin of electron (muon), and $g'_{e(\mu)}$ for the g factor of a bound electron (muon) in the muonium atom. The bound g factors are now known up to the fourth-order corrections [55] including the term of the order α^4, $\alpha^3 m_e/m_\mu$ and $\alpha^2 m_e/m_\mu$ and thus the relative uncertainty is essentially better than 10^{-8}. In particular, the result for the bound muon g factor reads [55][2]

$$g'_\mu = g_\mu^{(0)} \cdot \left\{ 1 - \frac{\alpha(Z\alpha)}{3} \left[1 - \frac{3}{2} \frac{m_e}{m_\mu} \right] \right.$$
$$\left. - \frac{\alpha(Z\alpha)(1+Z)}{2} \left(\frac{m_e}{m_\mu} \right)^2 + \frac{\alpha^2(Z\alpha)}{12\pi} \frac{m_e}{m_\mu} - \frac{97}{108} \alpha(Z\alpha)^3 \right\}, \quad (11)$$

where $g_\mu^{(0)} = 2 \cdot (1 + a_\mu)$ is the g factor of a free muon. Equation (10) has been applied [38,39] to determine the muon magnetic moment and muon mass by measuring the splitting of sublevels in the hyperfine structure of the $1s$ state in muonium in a homogeneous magnetic field. Their dependence on the magnetic field is given by the well known Breit-Rabi formula (see e.g. [56]). Since the magnetic field was calibrated via spin precession of the proton, the muon magnetic moment was measured in units of the proton magnetic moment, and muon-to-electron mass ratio was derived as

$$\frac{m_\mu}{m_e} = \frac{\mu_\mu}{\mu_p} \frac{\mu_p}{\mu_B} \frac{1}{1 + a_\mu}. \quad (12)$$

Results on the muon mass extracted from the Breit-Rabi formula are among the most accurate (see Fig. 9). A more precise value can only be derived from the muonium hyperfine structure after comparison of the experimental result with theoretical calculations. However, the latter is of less interest, since the most important application of the precise value of the muon-to-electron mass is to use it as an *input* for calculations of the muonium hyperfine structure while testing QED or determining the fine structure constants α. The adjusted CODATA result in Fig. 9 was extracted from the muonium hyperfine structure studies and in addition used some overoptimistic estimation of the theoretical uncertainty (see [36] for detail).

7 g Factor of a Bound Electron in a Hydrogen-Like Ion with Spinless Nucleus

In the case of an atom with a conventional nucleus (hydrogen, deuterium etc.) another notation is used and the expression for the Hamiltonian similar to eq. (10) can be applied. It can be used to test QED theory as well as to determine the electron-to-proton mass ratio. We underline that in contrast to most other tests it is possible to do both simultaneously because of a possibility to perform experiments with different ions.

The theoretical expression for the g factor of a bound electron can be presented in the form [3,58,59]

$$g'_e = 2 \cdot (1 + a_e + b), \quad (13)$$

[2] A misprint for the $\alpha^2(Z\alpha)m_e/m_\mu$ in [55] term is corrected here

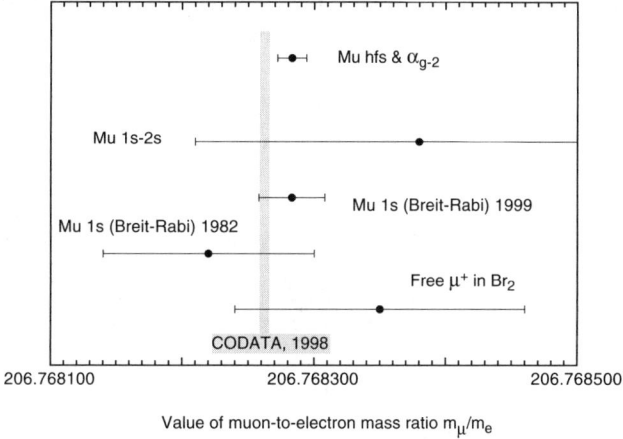

Value of muon-to-electron mass ratio m_μ/m_e

Fig. 9. The muon-to-electron mass ratio. The most accurate result obtained from comparison of the measured hyperfine interval in muonium [38] to the theoretical calculation [36] performed with $\alpha_{g-2}^{-1} = 137.035\,999\,58(52)$ [37]. The results derived from the Breit-Rabi sublevels are related to two experiments performed at LAMPF in 1982 [39] and 1999 [38]. The others are taken from the measurement of the $1s - 2s$ interval in muonium [57], precession of a free muon in bromine [40] and from the CODATA adjustment [10]

Table 6. The bound electron g factor in low-Z hydrogen-like ions with spinless nucleus

Ion	g
$^4\mathrm{He}^+$	$2.002\,177\,406\,7(1)$
$^{10}\mathrm{Be}^{3+}$	$2.001\,751\,574\,5(4)$
$^{12}\mathrm{C}^{5+}$	$2.001\,041\,590\,1(4)$
$^{16}\mathrm{O}^{7+}$	$2.000\,047\,020\,1(8)$
$^{18}\mathrm{O}^{7+}$	$2.000\,047\,021\,3(8)$

where the anomalous magnetic moment of a free electron $a_e = 0.001\,159\,652\,2$ [60,10] is known with good enough accuracy and b is the bound correction. The summary of the calculation of the bound corrections is presented in Table 6. The uncertainty of unknown two-loop contributions is taken from [61]. The calculation of the one-loop self-energy is different for different atoms. For lighter elements (helium, beryllium), it is obtained from [55] based on fitting data of [62], while for heavier ions we use the results of [63]. The other results are taken from [61] (for the one-loop vacuum polarization), [59] (for the nuclear correction and the electric part of the light-by-light scattering (Wichmann-Kroll) contribution), [64] (for the magnetic part of the light-by-light scattering contribution) and [65] (for the recoil effects).

Before comparing theory and experiment, let us shortly describe some details of the experiment. To determine a quantity like the g factor, one needs to measure some frequency at some known magnetic field B. It is clear that there is no way to directly determine magnetic field with a high accuracy. The conventional way is to measure two frequencies and to compare them. The frequencies measured in the GSI-Mainz experiment [68] are the ion cyclotron frequency

$$\omega_c = \frac{(Z-1)e}{M_i} B \tag{14}$$

and the Larmor spin precession frequency for a hydrogen-like ion with spinless nucleus

$$\omega_L = g_b \frac{e}{2m_e} B \ , \tag{15}$$

where M_i is the ion mass.

Combining them, one can obtain a result for the g factor of a bound electron

$$\frac{g_b}{2} = (Z-1) \frac{m_e}{M_i} \frac{\omega_L}{\omega_c} \tag{16}$$

or an electron-to-ion mass ratio

$$\frac{m_e}{M_i} = \frac{1}{Z-1} \frac{g_b}{2} \frac{\omega_c}{\omega_L} \ . \tag{17}$$

Today the most accurate value of m_e/M_i (without using experiments on the bound g factor) is based on a measurement of m_e/m_p realized in Penning trap [66] with a fractional uncertainty of 2 ppm. The accuracy of measurements of ω_c and ω_L as well as the calculation of g_b (as shown in [58]) are essentially better. That means that it is preferable to apply (17) to determine the electron-to-ion mass ratio [67]. Applying the theoretical value for the g factor of the bound electron and using experimental results for ω_c and ω_L in hydrogen-like carbon [68] and some auxiliary data related to the proton and ion masses, from [10], we arrive at the following values

$$\frac{m_p}{m_e} = 1\,836.152\,673\,1(10) \tag{18}$$

and

$$m_e = 0.000\,548\,579\,909\,29(31) \text{ u} \ , \tag{19}$$

which differ slightly from those in [67]. The present status of the determination of the electron-to-proton mass ratio is summarized in Fig. 10.

In [58] it was also suggested in addition to the determination of the electron mass to check theory by comparing the g factor for two different ions. In such a case the uncertainty related to m_e/M_i in (16) vanishes. Comparing the results for carbon [68] and oxygen [69], we find

$$g(^{12}\text{C}^{5+})/g(^{16}\text{O}^{7+}) = 1.000\,497\,273\,3(9) \tag{20}$$

to be compared to the experimental ratio

$$g(^{12}\text{C}^{5+})/g(^{16}\text{O}^{7+}) = 1.000\,497\,273\,1(15) \ . \tag{21}$$

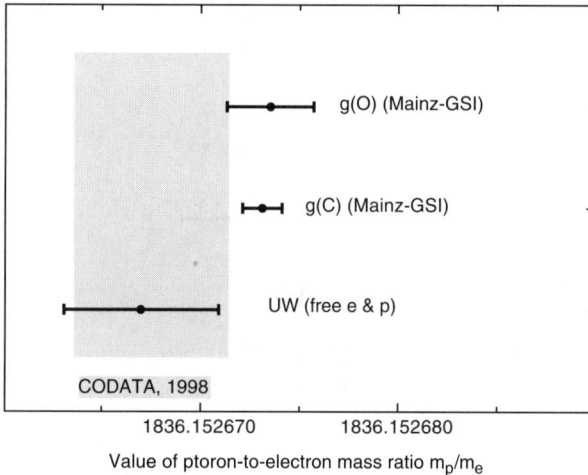

Fig. 10. The proton-to-electron mass ratio. The theory of the bound g factor is taken from Table 6, while the experimental data on the g factor in carbon and oxygen are from [68,69]. The Penning trap result from University of Washington is from [66]

Theory appears to be in fair agreement with experiment. In particular, this means that we have a reasonable estimate of uncalculated higher-order terms. Note, however, that for metrological applications it is preferable to study lower Z ions (hydrogen-like helium ($^4\text{He}^+$) and beryllium ($^{10}\text{Be}^{3+}$)) to eliminate these higher-order terms.

8   The Fine Structure Constant

The fine structure constant plays a basic role in QED tests. In atomic and particle physics there are several ways to determine its value. The results are summarized in Fig. 11. One method based on the muonium hyperfine interval was briefly discussed in Sect. 5. A value of the fine structure constant can also be extracted from the neutral-helium fine structure [70,71] and from the comparison of theory [37] and experiment [60] for the anomalous magnetic moment of electron (α_{g-2}). The latter value has been the most accurate one for a while and there was a long search for another competitive value. The second value (α_{Cs}) on the list of the most precise results for the fine structure constant is a result from recoil spectroscopy [72].

We would like to briefly consider the use and the importance of the recoil result for the determination of the fine structure constant. Absorbing and emitting a photon, an atom can gain some kinetic energy which can be determined as a shift of the emitted frequency in respect to the absorbed one (δf). A measurement of the frequency with high accuracy is the goal of the photon recoil experiment [72]. Combining the absorbed frequency and the shifted one, it is possible to determine a value of atomic mass (in [72] that was caesium) in fre-

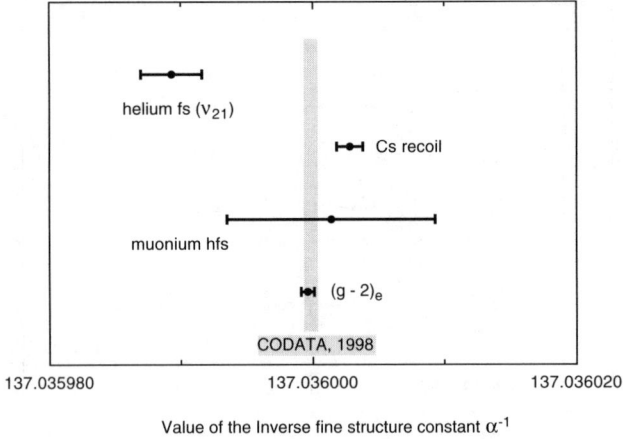

Fig. 11. The fine structure constant from atomic physics and QED

quency units, i.e. a value of $M_a c^2/h$. That may be compared to the Rydberg constant $Ry = \alpha^2 m_e c/2h$. The atomic mass is known very well in atomic units (or in units of the proton mass) [73], while the determination of electron mass in proper units is more complicated because of a different order of magnitude of the mass. The biggest uncertainty of the recoil photon value of α_{Cs} comes now from the experiment [72], while the electron mass is the second source.

The success of α_{Cs} determination was ascribed to the fact that α_{g-2} is a QED value being derived with the help of QED theory of the anomalous magnetic moment of electron, while the photon recoil result is free of QED. We would like to emphasize that the situation is not so simple and involvement of QED is not so important. It is more important that the uncertainty of α_{g-2} originates from understanding of the electron behaviour in the Penning trap and it dominates any QED uncertainty. For this reason, the value of α_{Cs} from m_p/m_e in the Penning trap [66] obtained by the same group as the one that determined the value of the anomalous magnetic moment of electron [60], can actually be correlated with α_{g-2}. The result

$$\alpha_{Cs}^{-1} = 137.036\,0002\,8(10) \tag{22}$$

presented in Fig. 11 is obtained using m_p/m_e from (18). The value of the proton-to-electron mass ratio found this way is free of the problems with an electron in the Penning trap, but some QED is involved. However, it is easy to realize that the QED uncertainty for the g factor of a bound electron and for the anomalous magnetic moment of a free electron are very different. The bound theory deals with simple Feynman diagrams but in Coulomb field and in particular to improve theory of the bound g factor, we need a better understanding of Coulomb effects for "simple" two-loop QED diagrams. In contrast, for the free electron no Coulomb field is involved, but a problem arises because of the four-loop diagrams. There is no correlation between these two calculations.

Table 7. Comparison of experiment and theory of hyperfine structure in hydrogen-like atoms. The numerical results are presented for the frequency E/h. In the D_{21} case the reference is given only for the $2s$ hyperfine interval

Atom	Experiment [kHz]	Theory [kHz]	Δ/σ	σ/E_F [ppm]
Hydrogen, D_{21}	49.13(15), [21,22]	48.953(3)	1.2	0.10
Hydrogen, D_{21}	48.53(23), [23]		-1.8	0.16
Hydrogen, D_{21}	49.13(40), [24]		0.4	0.28
Deuterium, D_{21}	11.16(16), [25]	11.312 5(5)	-1.0	0.49
$^3\mathrm{He}^+$ ion, D_{21}	-1 189.979(71), [26]	-1 190.072(63)	1.0	0.01
$^3\mathrm{He}^+$, D_{21}	-1 190.1(16), [27]		0.0	0.18
Muonium, $1s$	4 463 302.78(5)	4 463 302.88(55)	-0.18	0.11
Positronium, $1s$	203 389 100(740)	203 391 700(500)	-2.9	4.4
Positronium, $1s$	203 397 500(1600)		-2.5	8.2

9 Summary

To summarize QED tests related to hyperfine structure, we present in Table 7 the data related to hyperfine structure of the $1s$ state in positronium and muonium and to the D_{21} value in hydrogen, deuterium and helium-3 ion. The theory agrees with the experiment very well.

The precision physics of light simple atoms provides us with an opportunity to check higher-order effects of the perturbation theory. The highest-order terms important for comparison of theory and experiment are collected in Table 8. The uncertainty of the g factor of the bound electron in carbon and oxygen is related to $\alpha^2(Z\alpha)^4 m$ corrections in energy units, while for calcium the crucial order is $\alpha^2(Z\alpha)^6 m$.

Some of the corrections presented in Table 8 are completely known, some not. Many of them and in particular $\alpha(Z\alpha)^6 m^2/M^3$ and $(Z\alpha)^7 m^2/M^3$ for the hyperfine structure in muonium and helium ion, $\alpha^2(Z\alpha)^6 m$ for the Lamb shift in hydrogen and helium ion, $\alpha^7 m$ for positronium have been known in a so-called logarithmic approximation. In other words, only the terms with the highest power of "big" logarithms (e.g. $\ln(1/Z\alpha) \sim \ln(M/m) \sim 5$ in muonium) have been calculated. This program started for non-relativistic systems in [42] and was developed in [46,8,74,31,15]. By now even some non-leading logarithmic terms have been evaluated by several groups [43,75]. It seems that we have reached some numerical limit related to the logarithmic contribution and the calculation of the non-logarithmic terms will be much more complicated than anything else done before.

Twenty years ago, when I joined the QED team at Mendeleev Institute and started working on theory of simple atoms, experiment for most QED tests was considerably better than theory. Since that time several groups and independent scientists from Canada, Germany, Poland, Russia, Sweden and USA have been working in the field and moved theory to a dominant position. Today we are

Table 8. Comparison of QED theory and experiment: crucial orders of magnitude (see [2] for detail). Relativistic units in which $c = 1$ are used in the table

Value	Order
Hydrogen, deuterium (gross structure)	$\alpha(Z\alpha)^7 m$, $\alpha^2(Z\alpha)^6 m$
Hydrogen, deuterium (fine structure)	$\alpha(Z\alpha)^7 m$, $\alpha^2(Z\alpha)^6 m$
Hydrogen, deuterium (Lamb shift)	$\alpha(Z\alpha)^7 m$, $\alpha^2(Z\alpha)^6 m$
$^3\text{He}^+$ ion ($2s$ HFS)	$\alpha(Z\alpha)^7 m^2/M$, $\alpha(Z\alpha)^6 m^3/M^2$, $\alpha^2(Z\alpha)^6 m^2/M$, $(Z\alpha)^7 m^3/M^2$
$^4\text{He}^+$ ion (Lamb shift)	$\alpha(Z\alpha)^7 m$, $\alpha^2(Z\alpha)^6 m$
N^{6+} ion (fine structure)	$\alpha(Z\alpha)^7 m$, $\alpha^2(Z\alpha)^6 m$
Muonium ($1s$ HFS)	$(Z\alpha)^7 m^3/M^2$, $\alpha(Z\alpha)^6 m^3/M^2$, $\alpha(Z\alpha)^7 m^2/M$
Positronium ($1s$ HFS)	$\alpha^7 m$
Positronium (gross structure)	$\alpha^7 m$
Positronium (fine structure)	$\alpha^7 m$
Para-positronium (decay rate)	$\alpha^7 m$
Ortho-positronium (decay rate)	$\alpha^8 m$
Para-positronium (4γ branching)	$\alpha^8 m$
Ortho-positronium (5γ branching)	$\alpha^8 m$

looking forward to obtaining new experimental results to provide us with exciting data.

At the moment the ball is on the experimental side and the situation looks as if theorists should just wait. The theoretical progress may slow down because of no apparent strong motivation, but that would be very unfortunate. It is understood that some experimental progress is possible in near future with the experimental accuracy surpassing the theoretical one. And it is clear that it is extremely difficult to improve precision of theory significantly and we, theorists, have to start our work on this improvement now.

Acknowledgements

I am grateful to S.I. Eidelman, M. Fischer, T.W. Hänsch, E. Hessels, V.G. Ivanov, N. Kolachevsky, A.I. Milstein, P. Mohr, V.M. Shabaev, V.A. Shelyuto, and G. Werth for useful and stimulating discussions. This work was supported in part by the RFBR under grants ## 00-02-16718, 02-02-07027, 03-02-16843.

References

1. M.I. Eides, H. Grotch and V.A Shelyuto, Phys. Rep. **342**, 63 (2001)
2. S.G. Karshenboim, F.S. Pavone, F. Bassani, M. Inguscio and T.W. Hänsch: *Hydrogen atom: Precision physics of simple atomic systems* (Springer, Berlin, Heidelberg, 2001)

3. S.G. Karshenboim: In *Atomic Physics* **17** (AIP conference proceedings 551) Ed. by E. Arimondo et al. (AIP, 2001), p. 238

4. S.G. Karshenboim: In *Laser Physics at the Limits*, ed. by H. Figger, D. Meschede and C. Zimmermann (Springer-Verlag, Berlin, Heidelberg, 2001), p. 165

5. E.A. Hinds: In *The Spectrum of Atomic Hydrogen: Advances.* Ed. by. G.W. Series. (World Scientific, Singapore, 1988), p. 243

6. F.M. Pipkin: In *Quantum Electrodynamics.* Ed. by T. Kinoshita (World Scientific, Singapore, 1990), p. 479

7. F. Biraben, T.W. Hänsch, M. Fischer, M. Niering, R. Holzwarth, J. Reichert, Th. Udem, M. Weitz, B. de Beauvoir, C. Schwob, L. Jozefowski, L. Hilico, F. Nez, L. Julien, O. Acef, J.-J. Zondy, and A. Clairon: In [2], p. 17

8. S.G. Karshenboim: JETP **79**, 230 (1994)

9. S.G. Karshenboim: Z. Phys. D **39**, 109 (1997)

10. P.J. Mohr and B.N. Taylor: Rev. Mod. Phys. **72**, 351 (2000)

11. S.G. Karshenboim: Can. J. Phys. **77**, 241 (1999)

12. G.G. Simon, Ch. Schmitt, F. Borkowski and V.H. Walther: Nucl. Phys. A **333**, 381 (1980)

13. R. Pohl, F. Biraben, C.A.N. Conde, C. Donche-Gay, T.W. Hänsch, F. J. Hartmann, P. Hauser, V.W. Hughes, O. Huot, P. Indelicato, P. Knowles, F. Kottmann, Y.-W. Liu, V. E. Markushin, F. Mulhauser, F. Nez, C. Petitjean, P. Rabinowitz, J.M.F. dos Santos, L.A. Schaller, H. Schneuwly, W. Schott, D. Taqqu, and J.F.C.A. Veloso: In [2], p. 454

14. I.B. Khriplovich, A. I. Milstein, and S.S. Petrosyan: Phys. Lett. B **366**, 13 (1996), JETP **82**, 616 (1996)

15. S.G. Karshenboim and V.G. Ivanov: Phys. Lett. B **524**, 259 (2002); Euro. Phys. J. D **19**, 13 (2002)

16. S.G. Karshenboim: Can. J. Phys. **78**, 639 (2000)

17. H. Hellwig, R.F.C. Vessot, M.W. Levine, P.W. Zitzewitz, D.W. Allan, and D.J. Glaze: IEEE Trans. IM **19**, 200 (1970);
P.W. Zitzewitz, E.E. Uzgiris, and N.F. Ramsey: Rev. Sci. Instr. **41**, 81 (1970);
D. Morris: Metrologia **7**, 162 (1971);
L. Essen, R. W. Donaldson, E. G. Hope and M.J. Bangham: Metrologia **9**, 128 (1973);
J. Vanier and R. Larouche: Metrologia **14**, 31 (1976);
Y.M. Cheng, Y.L. Hua, C.B. Chen, J.H. Gao and W. Shen: IEEE Trans. IM **29**, 316 (1980);
P. Petit, M. Desaintfuscien and C. Audoin: Metrologia **16**, 7 (1980)

18. D.J. Wineland and N. F. Ramsey: Phys. Rev. **5**, 821 (1972)

19. B.S. Mathur, S.B. Crampton, D. Kleppner and N.F. Ramsey: Phys. Rev. **158**, 14 (1967)

20. H.A. Schluessler, E.N. Forton and H.G. Dehmelt: Phys. Rev. **187**, 5 (1969)

21. M. Fischer, N. Kolachevsky, S G. Karshenboim and T.W. Hänsch: eprint physics/0305073

22. M. Fischer, N. Kolachevsky, S.G. Karshenboim and T.W. Hänsch: Can. J. Phys. **80**, 1225 (2002)

23. N.E. Rothery and E. A. Hessels: Phys. Rev. A **61**, 044501 (2000)

24. J.W. Heberle, H.A. Reich and P. Kush: Phys. Rev. **101**, 612 (1956)

25. H.A. Reich, J.W. Heberle, and P. Kush: Phys. Rev. **104**, 1585 (1956)

26. M.H. Prior and E.C. Wang: Phys. Rev. A **16**, 6 (1977)

27. R. Novick and D.E. Commins: Phys. Rev. **111**, 822 (1958)

28. D. Zwanziger: Phys. Rev. **121**, 1128 (1961)
29. M. Sternheim: Phys. Rev. **130**, 211 (1963)
30. P. Mohr: unpublished. Quoted according to [26]
31. S.G. Karshenboim: In [2], p. 335
32. V.A. Yerokhin and V.M. Shabaev: Phys. Rev. A **64**, 012506 (2001)
33. G. Breit: Phys. Rev. **35**, 1477 (1930)
34. K. Jungmann: In [2], p. 81
35. R.S. Conti, R.S. Vallery, D.W. Gidley, J.J. Engbrecht, M. Skalsey, and P.W. Zitzewitz: In [2], p. 103
36. A. Czarnecki, S.I. Eidelman and S.G. Karshenboim: Phys. Rev. D**65**, 053004 (2002)
37. V.W. Hughes and T. Kinoshita: Rev. Mod. Phys. **71** (1999) S133;
    T. Kinoshita: In [2], p. 157
38. W. Liu, M.G. Boshier, S. Dhawan, O. van Dyck, P. Egan, X. Fei, M. G. Perdekamp, V.W. Hughes, M. Janousch, K. Jungmann, D. Kawall, F.G. Mariam, C. Pillai, R. Prigl, G. zu Putlitz, I. Reinhard, W. Schwarz, P.A. Thompson, and K.A. Woodle: Phys. Rev. Lett. **82**, 711 (1999)
39. F.G. Mariam, W. Beer, P.R. Bolton, P.O. Egan, C.J. Gardner, V.W. Hughes, D C. Lu, P.A. Souder, H. Orth, J. Vetter, U. Moser, and G. zu Putlitz: Phys. Rev. Lett. **49** (1982) 993
40. E. Klempt, R. Schulze, H. Wolf, M. Camani, F. Gygax, W. Rüegg, A. Schenck, and H. Schilling: Phys. Rev. D **25** (1982) 652
41. G.S. Adkins, R.N. Fell, and P. Mitrikov: **79**, 3383 (1997);
    A.H. Hoang, P. Labelle, and S M. Zebarjad: **79**, 3387 (1997)
42. S.G. Karshenboim: JETP **76**, 541 (1993)
43. K. Melnikov and A. Yelkhovsky: Phys. Rev. Lett. **86** (2001) 1498;
    R. Hill: Phys. Rev. Lett. **86** (2001) 3280;
    B. Kniehl and A.A. Penin: Phys. Rev. Lett. **85**, 5094 (2000)
44. S. G. Karshenboim: Appl. Surf. Sci. **194**, 307 (2002)
45. G.W. Bennett, B. Bousquet, H.N. Brown, G. Bunce, R. M. Carey, P. Cushman, G.T. Danby, P.T. Debevec, M. Deile, H. Deng, W. Deninger, S.K. Dhawan, V.P. Druzhinin, L. Duong, E. Efstathiadis, F.J.M. Farley, G.V. Fedotovich, S. Giron, F.E. Gray, D. Grigoriev, M. Grosse-Perdekamp, A. Grossmann, M.F. Hare, D.W. Hertzog, X. Huang, V.W. Hughes, M. Iwasaki, K. Jungmann, D. Kawall, B.I. Khazin, J. Kindem, F. Krienen, I. Kronkvist, A. Lam, R. Larsen, Y.Y. Lee, I. Logashenko, R. McNabb, W. Meng, J. Mi, J.P. Miller, W.M. Morse, D. Nikas, C.J.G. Onderwater, Y. Orlov, C.S. Özben, J.M. Paley, Q. Peng, C.C. Polly, J. Pretz, R. Prigl, G. zu Putlitz, T. Qian, S.I. Redin, O. Rind, B.L. Roberts, N. Ryskulov, P. Shagin, Y.K. Semertzidis, Yu.M. Shatunov, E.P. Sichtermann, E. Solodov, M. Sossong, A. Steinmetz, L.R. Sulak, A. Trofimov, D. Urner, P. von Walter, D. Warburton, and A. Yamamoto: Phys. Rev. Lett. **89**, 101804 (2002);
    S. Redin, R.M. Carey, E. Efstathiadis, M.F. Hare, X. Huang, F. Krinen, A. Lam, J.P. Miller, J. Paley, Q. Peng, O. Rind, B.L. Roberts, L.R. Sulak, A. Trofimov, G.W. Bennett, H.N. Brown, G. Bunce, G.T. Danby, R. Larsen, Y Y. Lee, W. Meng, J. Mi, W.M. Morse, D. Nikas, C. Ozben, R. Prigl, Y.K. Semertzidis, D. Warburton, V.P. Druzhinin, G.V. Fedotovich, D. Grigoriev, B.I. Khazin, I.B. Logashenko, N.M. Ryskulov, Yu.M. Shatunov, E.P. Solodov, Yu.F. Orlov, D. Winn, A. Grossmann, K. Jungmann, G. zu Putlitz, P. von Walter, P.T. Debevec, W. Deninger, F. Gray, D.W. Hertzog, C.J.G. Onderwater, C. Polly, S. Sedykh, M. Sossong, D. Urner, A. Yamamoto, B. Bousquet, P. Cushman, L. Duong, S. Giron, J. Kindem, I. Kronkvist, R. McNabb, T. Qian, P. Shagin, C. Timmermans, D. Zimmerman,

M. Iwasaki, M. Kawamura, M. Deile, H. Deng, S.K. Dhawan, F.J.M. Farley, M. Grosse-Perdekamp, V.W. Hughes, D. Kawall, J. Pretz, E.P. Sichtermann, and A. Steinmetz: Can. J. Phys. **80**, 1355 (2002);
S. Redin, G.W. Bennett, B. Bousquet, H.N. Brown, G. Bunce, R.M. Carey, P. Cushman, G T. Danby, P.T. Debevec, M. Deile, H. Deng, W. Deninger, S.K. Dhawan, V.P. Druzhinin, L. Duong, E. Efstathiadis, F.J.M. Farley, G.V. Fedotovich, S. Giron, F. Gray, D. Grigoriev, M. Grosse-Perdekamp, A. Grossmann, M.F. Hare, D.W. Hertzog, X. Huang, V.W. Hughes, M. Iwasaki, K. Jungmann, D. Kawall, M. Kawamura, B.I. Khazin, J. Kindem, F. Krinen, I. Kronkvist, A. Lam, R. Larsen, Y.Y. Lee, I.B. Logashenko, R. McNabb, W. Meng, J. Mi, J.P. Miller, W.M. Morse, D. Nikas, C.J. G. Onderwater, Yu.F. Orlov, C. Ozben, J. Paley, Q. Peng, J. Pretz, R. Prigl, G. zu Putlitz, T. Qian, O. Rind, B.L. Roberts, N.M. Ryskulov, P. Shagin, S. Sedykh, Y.K. Semertzidis, Yu.M. Shatunov, E.P. Solodov, E.P. Sichtermann, M. Sossong, A. Steinmetz, L.R. Sulak, C. Timmermans, A. Trofimov, D. Urner, P. von Walter, D. Warburton, D. Winn, A. Yamamoto, and D. Zimmerman: In *This book*, pp. 163–174
46. S.G. Karshenboim: Z. Phys. D **36**, 11 (1996)
47. S.G. Karshenboim and V.A. Shelyuto: Phys. Lett. B **517**, 32 (2001)
48. S.I. Eidelman, S.G. Karshenboim and V.A. Shelyuto: Can. J. Phys. **80**, 1297 (2002)
49. S. Geer: Phys. Rev. D **57**, 6989 (1998); D **59**, 039903 (E) (1998);
B. Autin, A. Blondel and J. Ellis (eds.): *Prospective study of muon storage rings at CERN*, Report CERN-99-02
50. J.R. Sapirstein, E.A. Terray, and D.R. Yennie: Phys. Rev. D **29**, 2290 (1984)
51. A. Karimkhodzhaev and R.N. Faustov: Sov. J. Nucl. Phys. **53**, 626 (1991)
52. R.N. Faustov, A. Karimkhodzhaev and A.P. Martynenko: Phys. Rev. A **59**, 2498 (1999)
53. M.W. Ritter, P.O. Egan, V.W. Hughes and K.A. Woodle: Phys. Rev. A **30**, 1331 (1984)
54. A.P. Mills, Jr., and G.H. Bearman: Phys. Rev. Lett. **34**, 246 (1975);
A. P. Mills, Jr.: Phys. Rev. A **27**, 262 (1983)
55. S.G. Karshenboim and V.G. Ivanov: Can. J. Phys. **80**, 1305 (2002)
56. H.A. Bethe and E.E. Salpeter: *Quantum Mechanics of One- and Two-electon Atoms* (Plenum, NY, 1977)
57. F. Maas, B. Braun, H. Geerds, K. Jungmann, B.E. Matthias, G. zu Putlitz, I. Reinhard, W. Schwarz, L. Williams, L. Zhang, P.E.G. Baird, P.G.H. Sandars, G. Woodman, G.H. Eaton, P. Matousek, T. Toner, M. Towrie, J.R.M. Barr, A.I. Ferguson, M.A. Persaud, E. Riis, D. Berkeland, M. Boshier and V. W. Hughes: Phys. Lett. A **187**, 247 (1994)
58. S.G. Karshenboim: In [2], p. 651
59. S.G. Karshenboim: Phys. Lett. A **266**, 380 (2000)
60. R.S. Van Dyck Jr., P.B. Schwinberg, and H.G. Dehmelt: Phys. Rev. Lett. **59**, 26 (1987)
61. S.G. Karshenboim, V.G. Ivanov and V M. Shabaev: JETP **93**, 477 (2001); Can. J. Phys. **79**, 81 (2001)
62. T. Beier, I. Lindgren, H. Persson, S. Salomonson, and P. Sunnergren: Phys. Rev. A **62**, 032510 (2000), Hyp. Int. **127**, 339 (2000)
63. V.A. Yerokhin, P. Indelicato and V.M. Shabaev: Phys. Rev. Lett. **89**, 143001 (2002)
64. S.G. Karshenboim and A.I. Milstein: Physics Letters B **549**, 321 (2002)
65. V.M. Shabaev and V.A. Yerokhin: Phys. Rev. Lett. **88**, 091801 (2002)
66. D.L. Farnham, R.S. Van Dyck, Jr., and P.B. Schwinberg: Phys. Rev. Lett. **75**, 3598 (1995)

67. T. Beier, H. Häffner, N. Hermanspahn, S.G. Karshenboim, H.-J. Kluge, W. Quint, S. Stahl, J. Verdú, and G. Werth: Phys. Rev. Lett. **88**, 011603 (2002)
68. H. Häffner, T. Beier, N. Hermanspahn, H.-J. Kluge, W. Quint, S. Stahl, J. Verdú, and G. Werth: Phys. Rev. Lett. **85**, 5308 (2000);
    G. Werth, H. Häffner, N. Hermanspahn, H.-J. Kluge, W. Quint, and J. Verdú: In [2], p. 204
69. J.L. Verdú, S. Djekic, T. Valenzuela, H. Häffner, W. Quint, H.J. Kluge, and G. Werth: Can. J. Phys. **80**, 1233 (2002);
    G. Werth: *private communication*
70. G.W.F. Drake: Can. J. Phys. **80**, 1195 (2002);
    K. Pachucki and J. Sapirstein: J. Phys. B: At. Mol. Opt. Phys. **35**, 1783 (2002)
71. M.C. George, L.D. Lombardi, and E.A. Hessels: Phys. Rev. Lett. **87**, 173002 (2001)
72. A. Wicht, J.M. Hensley, E. Sarajlic, and S. Chu: In *Proceedings of the 6th Symposium Frequency Standards and Metrology*, ed. by P. Gill (World Sci., 2002) p.193
73. S. Rainville, J.K. Thompson, and D.E. Pritchard: Can. J. Phys. **80**, 1329 (2002), In *This book*, pp. 177–197
74. K. Pachucki and S.G. Karshenboim: Phys. Rev. A **60**, 2792 (1999);
    K. Melnikov and A. Yelkhovsky: Phys. Lett. B **458**, 143 (1999)
75. K. Pachucki: Phys. Rev. A **63**, 042053 (2001)

Resent Results and Current Status of the Muon (g–2) Experiment at BNL

S.I. Redin[3,12], G.W. Bennett[2], B. Bousquet[10], H.N. Brown[2], G. Bunce[2], R.M. Carey[1], P. Cushman[10], G.T. Danby[2], P.T. Debevec[8], M. Deile[12], H. Deng[12], W. Deninger[8], S.K. Dhawan[12], V.P. Druzhinin[3], L. Duong[10], E. Efstathiadis[1], F.J.M. Farley[12], G.V. Fedotovich[3], S. Giron[10], F. Gray[8], D. Grigoriev[3], M. Grosse-Perdekamp[12], A. Grossmann[7], M.F. Hare[1], D.W. Hertzog[8], X. Huang[1], V.W. Hughes[12], M. Iwasaki[11], K. Jungmann[6], D. Kawall[12], M. Kawamura[11], B.I. Khazin[3], J. Kindem[10], F. Krinen[1], I. Kronkvist[10], A. Lam[1], R. Larsen[2], Y.Y. Lee[2], I.B. Logashenko[3], R. McNabb[10], W. Meng[2], J. Mi[2], J.P. Miller[1], W.M. Morse[2], D. Nikas[2], C.J.G. Onderwater[8], Yu.F. Orlov[4], C. Ozben[2], J. Paley[1], Q. Peng[1], J. Pretz[12], R. Prigl[2], G. zu Putlitz[7], T. Qian[10], O. Rind[1], B.L. Roberts[1], N.M. Ryskulov[3], P. Shagin[10], S. Sedykh[8], Y.K. Semertzidis[2], Yu.M. Shatunov[3], E.P. Solodov[3], E.P. Sichtermann[12], M. Sossong[8], A. Steinmetz[12], L.R. Sulak[1], C. Timmermans[10], A. Trofimov[1], D. Urner[8], P. von Walter[7], D. Warburton[2], D. Winn[5], A. Yamamoto[9], and D. Zimmerman[10]

[1] Department of Physics, Boston University, Boston, MA 02215, USA
[2] Brookhaven National Laboratory, Upton, NY 11973, USA
[3] Budker Institute of Nuclear Physics, Novosibirsk, Russia
[4] Newman Laboratory, Cornell University, Ithaca, NY 14853, USA
[5] Fairfield University, Fairfield, CT 06430, USA
[6] Kernfysisch Versneller Instituut, Rijksuniversiteit Groningen, KVI Zernikelaan 25, Groningen, NL 9747 AA, The Netherlands
[7] Physikalisches Institut der Universität Heidelberg, 69120 Heidelberg, Germany
[8] Department of Physics, University of Illinois at Urbana-Champaign, Urbana, IL 55455, USA
[9] KEK, High Energy Accelerator Research Organization, Tsukuba, Ibaraki 305-0801, Japan
[10] Department of Physics, University of Minnesota, Minneapolis, MN 55455, USA
[11] Tokyo Institute of Technology, Tokyo, Japan
[12] Physics Department, Yale University, New Haven, CT 06520, USA

Abstract. The measurement of the (g–2) value of leptons provides a unique test of theory since it is the only quantity (unlike charge and mass) calculable in the framework of the Standard Model of elementary particles. The muon (g–2) experiment E821 is currently in progress at Brookhaven National Laboratory. Four data taking runs for positive muons and one run for negative muons were successfully accomplished in 1997–2000 and 2001, respectively. Results of the 1997–2000 runs have been published, thus completing our experiment for μ^+. Data analysis for the 2001 run for μ^- is currently in progress. To provide measurement of $a_{\mu^-} = \frac{1}{2}(g-2)_{\mu^-}$ at the same level of accuracy as for $a_{\mu^+} = \frac{1}{2}(g-2)_{\mu^+}$, we need to have one more data taking run.

1   Introduction

Historically, the study of magnetic moments has played an important role in our understanding of subatomic physics. Precision measurements of the electron magnetic moment, together with the Lamb shift, led to the development of moderm quantum electrodynamics with its renormalization procedure, which is a corner-stone of modern high energy physics. The magnetic moment of a particle is related to its intrinsic spin via the gyromagnetic ratio,

$$\boldsymbol{\mu} = g \, \frac{e\hbar}{2mc} \, \boldsymbol{S} \, . \tag{1}$$

For a lepton the Dirac theory predicts $g = 2$. A deviation from $g = 2$, the so called *anomaly*, also referred to as the *anomalous magnetic moment* (presented in the unit of the Bohr magneton of the lepton), is defined as $a = \frac{1}{2}(g-2)$ and arises from radiative corrections and in principle can be calculated from quantum field theory. In the framework of the Standard Model a is a sum of quantum electrodynamics, hadronic and weak contributions,

$$a = a(\mathrm{QED}) + a(\mathrm{had}) + a(\mathrm{weak}) \, . \tag{2}$$

A precision measurement of a_e and a_μ probes the short–distance structure of the theory and hence provides a stringent test of the Standard Model or, alternatively, a search for New Physics. The contribution from heavy particles and high energy processes to a_μ is usually bigger than that to a_e by a factor of $\sim m_\mu^2/m_e^2 = 4 \cdot 10^4$, and hence a_μ is more sensitive to New Physics than a_e. In turn, a_e is almost completely determined by Quantum Electrodynamics corrections and therefore has small theoretical uncertainty. With a_e having been measured in experiment [1] with relative precision 4×10^{-9}, one might equate theoretical and experimental values and thus obtain the most precise value for the fine structure constant $\alpha(a_e)$. This $\alpha(a_e)$ is in good agreement with current results of direct measurements and is used for evaluation of $a_\mu(\mathrm{QED})$.

Previous measurement of a_μ at CERN [2] confirmed theoretically predicted contributions from QED and strong interactions with experimental precision of 7.2 ppm, but was not accurate enough to see contribution from weak interactions, which is about 1.3 ppm in relative units. The goal of the current BNL muon (g–2) experiment is to lower the experimental error to 0.35 ppm (a factor of ~20 improvement), which would allow us to see the contribution of weak interactions at the level of 3–4 standard deviations and search further for New Physics.

Four data taking runs for positive muons and one run for negative muons were successfully accomplished at BNL in 1997–2000 and 2001, respectively. Results of the 1997–2000 runs have been published in [3–6], thus completing our experiment for μ^+. We discuss these results in Sects. 4 and 5.

Data analysis for the 2001 run for μ^- is currently in progress. To provide measurement of $a_{\mu^-} = \frac{1}{2}(g-2)_{\mu^-}$ at the same level of accuracy as for $a_{\mu^+} = \frac{1}{2}(g-2)_{\mu^+}$, we need to have one more data taking run.

2   Muon (g–2) Experiment E821 at BNL

The principle of the BNL experiment is based on the spin motion of polarized muons in a storage ring and is the same as that for the most recent CERN experiment [2]. In a uniform magnetic field B the spin precesses with an angular frequency ω_s which is greater than the orbital cyclotron frequency ω_c by ω_a, which is the (g–2) precession frequency:

$$\omega_a = \omega_s - \omega_c = \left[g\,\frac{eB}{2mc} + (1-\gamma)\frac{eB}{\gamma mc} \right] - \frac{eB}{\gamma mc} = a_\mu \frac{eB}{mc}.$$

In order to retain muons in the storage ring vertical focusing is provided by electrostatic quadrupoles. With both magnetic and electric fields present, the expression for ω_a becomes

$$\boldsymbol{\omega}_a = \frac{e}{mc}\left[a_\mu \boldsymbol{B} - (a_\mu - \frac{1}{\gamma^2 - 1})\boldsymbol{\beta} \times \boldsymbol{E} \right]. \tag{3}$$

For a unique value of muon energy $E_\mu = 3.096$ GeV ($\gamma = 29.3$), the electric field does not contribute to ω_a. Choosing this "magic energy" allows us to separate the functions of the fields: the homogeneous magnetic field B determines the muon spin precession and electrostatic quadrupoles provide vertical focusing of the muon beam.

The (g–2) frequency ω_a appears as a modulation of the muon decay electron spectrum. Electrons from muon decay are detected with calorimeters and their times of arrival are accurately measured. An accurate determination of a_μ requires an accurate measurement of ω_a and of B. Magnetic field is measured by precession frequency ω_p of proton spin in a water sample inside NMR probe and hence a_μ can be evaluated in terms of ω_a to ω_p ratio and ratio of magnetic moments of muon and proton:

$$\frac{\mu_\mu}{\mu_p} = \frac{\omega_s(\text{rest})}{\omega_p} = g\,\frac{eB}{2\,mc\,\omega_p} = g\,\frac{\omega_a}{2\,a_\mu\,\omega_p} = \frac{1 + a_\mu}{a_\mu}\,\frac{\omega_a}{\omega_p}\,, \tag{4}$$

thus

$$a_\mu = \frac{\omega_a/\omega_p}{\mu_\mu/\mu_p - \omega_a/\omega_p}\,, \tag{5}$$

where $\mu_\mu/\mu_p = 3.183\,345\,39\,(10)\,(0.03\text{ ppm})$, obtained in [7].

The increased precision in our experiment is possible principally because of the high proton beam intensity of the AGS, which is about 200 times that available at CERN, where the dominant error was statistical. Our secondary beam line provides either π or μ beams for injection into the storage ring. With pion injection (explored in CERN and in our first data taking run in 1997 [3]), a small fraction of the muons from π decays is captured in the storage ring, while with muon injection the muon beam is injected directly from the beam line and a higher intensity of stored muons is provided. Direct muon injection for muon (g–2) measurements was introduced for the first time in our experiment.

Fig. 1. Muon storage ring at Brookhaven National Laboratory

It is provided by the fast kicker being comissioned prior to our 1998 data taking run [4].

For the first 30 μs after either type of injection, the voltage, applied to the quadrupoles, is made asymmetrical in order to shift muon orbit and thus get rid of muons at the edges of storage volume, i.e. those which otherwise would be uncontrollably lost later, at the data recording time, and thus influence the decay electron spectrum.

The BNL (g–2) storage ring is shown in Fig. 1. Twenty four detector stations [8] detect decay electrons on the inside of the ring. They handle high rates and provide precise time measurements with systematic errors less than 20 ps. Each station consists of an electromagnetic calorimeter and a horizontal array of five scintillator paddles on the front face of the calorimeter. The calorimeters are made of scintillator fibers embedded in lead. The radiation length of the calorimeter is about 1 cm, energy resolution is $10-13\%/\sqrt{E(\text{GeV})}$. Scintillator paddles provide additional time measurement and rough information about the vertical coordinate of the point, where the electron enters the calorimeter.

Figure 2 shows time distribution of 0.95 billion decay positrons for the 1999 data taking run [5]. It is basically an exponentially decaying sine wave with period being $T_a = 2\pi/\omega_a$. The data are fitted with an appropriate fit function and thus ω_a and its statistical error are obtained.

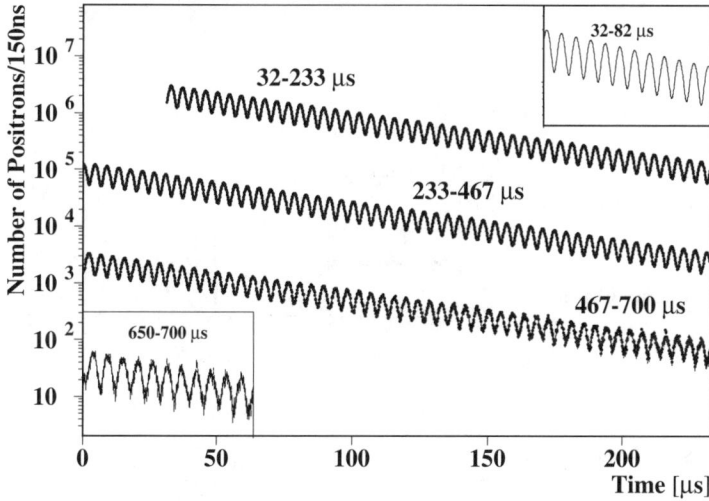

Fig. 2. Time distribution of a sample of high energy decay positrons

3   Magnetic Field Measurement and Control

The principal equipment for the experiment is the superferric storage ring [9]. To achieve the desired precision we must know B averaged over the muon storage volume at 0.1 ppm level. Hence the requirements on the field homogeneity and stability are very stringent.

The BNL (g–2) magnet provides a magnetic field of about 1.45 T over the muon storage region, which has a toroidal shape with the radius of the central orbit 711.2 cm and cross sectional diameter 9 cm. The cross section of the muon storage ring is shown in Fig. 3. The magnet is C-shaped to allow the decay electrons be observed inside the ring.

To measure and control the magnetic field with great accuracy, a pulsed NMR system has been developed [10]. The major part of the system consists of 17 NMR probes mounted on a beam tube trolley as shown in Fig. 4. They are used for the magnetic field mapping during the data taking runs, without breaking the vacuum. The measurements with the beam tube trolley are taken every 48 to 72 hours. Each field map is made up of about 6000 readings for each of the 17 NMR probes. Between these measurements the drift of the magnetic field is monitored by a set of about 150 high quality NMR probes out of 366 fixed probes embedded in the walls of the vacuum chamber as shown in Fig. 3, in 72 azimuthal locations. Some 36 fixed NMR probes are used for the feedback stabilization of magnet's power supply.

Fixed probes are calibrated by beam tube trolley probes during each field mapping. At the end of the run, all of the 17 trolley NMR probes are calibrated against a single calibration probe [10]. The latter, in turn, is calibrated with a standard NMR probe [11], which was constructed to measure the NMR frequency

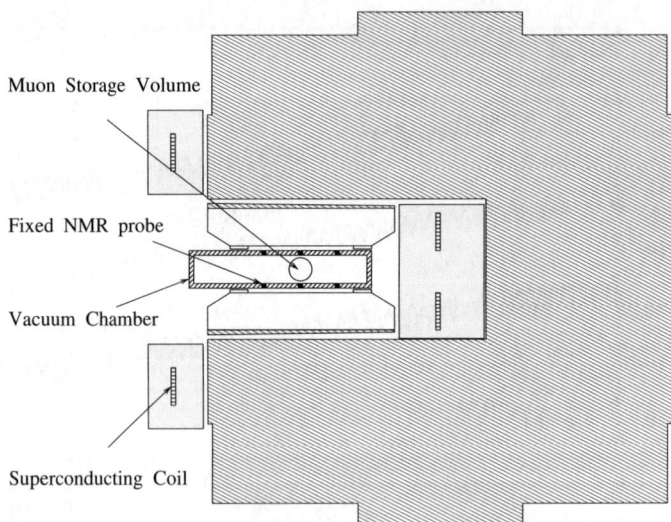

Fig. 3. Cross section of magnet and fixed NMR probes mounted in the vacuum chamber wall

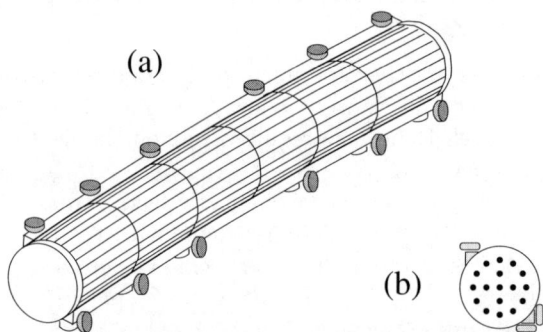

Fig. 4. Beam tube trolley (a) and scheme of location of 17 NMR probes inside it (b)

of protons in a spherical sample of pure water with a systematic uncertainty of 0.034 ppm.

Fig. 5 shows azimuthally averaged magnetic field across the muon storage region for 1999 (left) and 2000 (right) runs. Significant improvement of homogeneity of magnetic field was achieved by replacement of inflector magnet. This, and improved calibration procedure, allowed to reduce systematic error for ω_p from 0.4 ppm in 1999 to 0.24 ppm in 2000.

Two largely independent analyses of ω_p were made using different selection of fixed NMR probes for the field extrapolation between the beam tube trolley field mappings. Their results were found to agree to within 0.05 ppm. The final value for 2000 data is $\omega_p/(2\pi) = 61\ 791\ 595 \pm 15$ Hz (0.2 ppm).

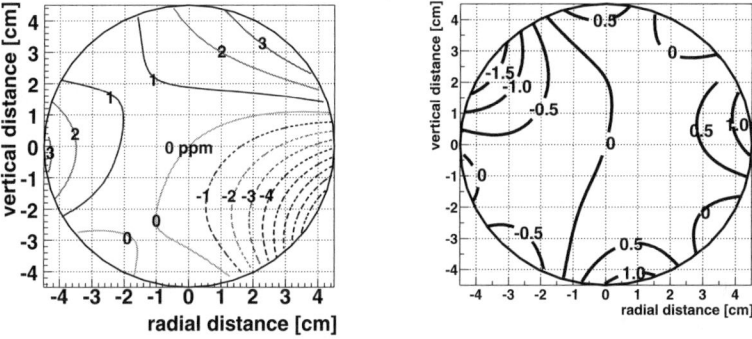

Fig. 5. Distribution of azimuthally averaged magnetic field across the muon storage region. One ppm and half ppm contours are shown with respect to a central field for 1999 (left) and for 2000 (right) data, respectively

4   Data Analysis and Results

In our 2000 data taking run we have observed 4 billion decay positrons. The time spectrum of decay positrons is basically the exponentially decaying sine wave:

$$f(t) = N_o \, e^{-t/\tau} \, [1 + A_a \sin(\omega_a t + \phi_a)] \tag{6}$$

with period $T_a = 2\pi/\omega_a = 4.37\,\mu s$ and lifetime $\tau = \gamma\tau_o = 64.4\,\mu s$. However, the spectrum is slightly distorted by (1) pulses overlapping in time (pileup), (2) muon losses and (3) coherent betatron oscillations (CBO).

(1) The pileup is proportional to e^+ rate squared and its level is $\sim 1\%$ at $30\,\mu s$ after the muon injection. Pileup contribution can be estimated and subsequently subtracted from the data (pileup subtraction) as described in [5].

(2) To monitor muon loss, we have implemented several muon loss monitors. Each of these monitors counts time coincidence of signals from the front scintillator paddles of three adjacent detector stations (triple coincidence). Since any other charged particle but muon is not able to pass through the material of two calorimeters, the loss monitor readout indicates the rate of muon losses from the beam as a function of time. The rate of muon losses was found to be $\sim 0.6\%$ at $30\,\mu s$.

(3) In the direct muon injection, due to inflector geometry and imperfect kick, the stored muon beam does not fill the available phase space uniformly. In general, the center of gravity of the muon distribution in the horizontal (radial) direction is off the center of the storage volume and shows betatron oscillations around the center with a period of $\nu_x^{-1} = 1.07$ in units of cyclotron period $T_c = 0.15\,\mu s$. An observer located at position of a particular detector system will see radial muon distribution to be replicated every $|1 - 1.07|^{-1} \approx 14^{th}$ turn. Since the acceptance of a detector station varies with the radial position of the decaying muons and the momentum of produced positrons, the time and energy spectra of positrons will be modulated with a period of $\sim 14\,T_c$, which is close

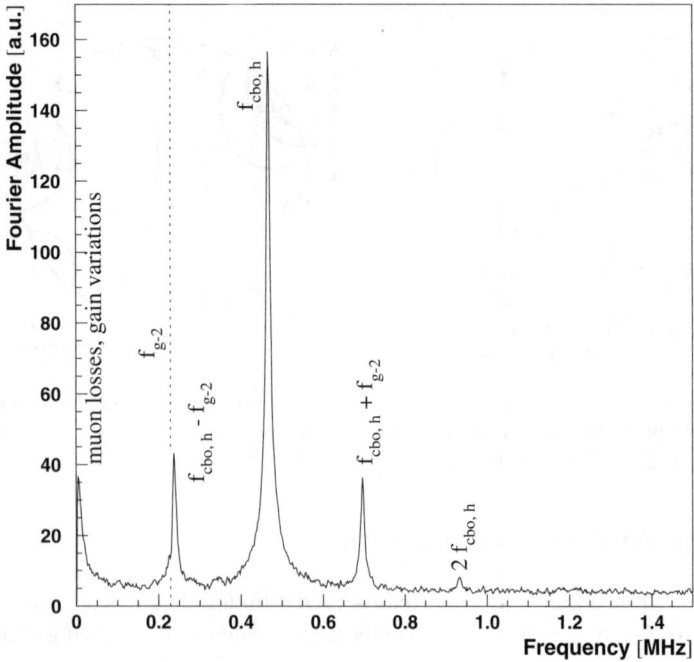

Fig. 6. Fourier spectrum of residual

to one half of period of (g–2) oscillations $T_a = (4.37\,\mu s)/(0.15\,\mu s)\,T_c = 29\,T_c$. Figure 6 shows the Fourier spectrum of the residual, which is the difference between the positron time distribution and the 5-parameter fit function (6). The strong peak at about 0.47 MHz is due to CBO modulation of decay positron rate (parameter N_o in (6)). Peaks from the second CBO harmonic and sidebands $\omega_{cbo} \pm \omega_a$ are also clearly seen. CBO modulation of energy spectrum of positrons causes modulation of asymmetry A_a and phase ϕ_a in (6) and, as a result, unequal magnitudes of the CBO sidebands in Fig. 6.

Four independent analyses of ω_a have been done by groups in Yale University, BNL, University of Illinois and University of Minnesota. In the first analysis the pileup subtracted time spectrum of all positrons with energies greater than 2 GeV and arrival times in the range 49–600 μs after muon injection is considered. The effects of horizontal CBO have been incorporated in the fitting function as modulations of the observed positron rate N_o and asymmetry A_a. The CBO modulation envelope was determined from the data by partial Fourier integration. Muon losses were constrained to the shape of the measured losses and their scale was included in set of the fit parameters.

The second analysis was similar to the first one with slightly different data selection, fit start times and set of fit parameters. Alternative methods were used to estimate systematic uncertainties. Notably, in one of them to study the systematic effects of CBO, the positron time spectrum was strobed at fixed

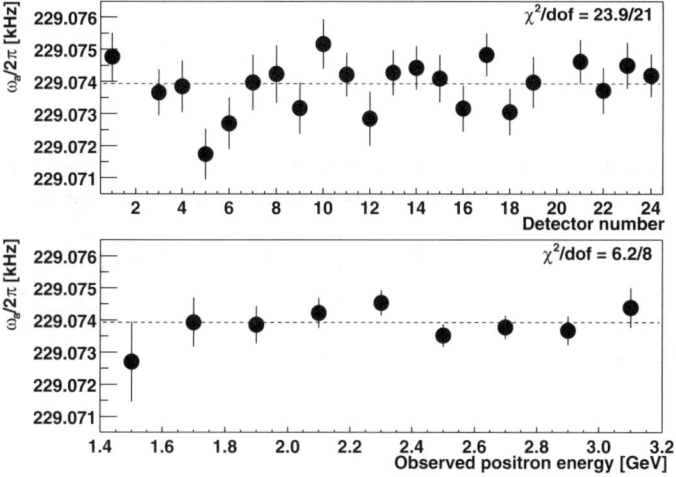

Fig. 7. Frequency ω_a versus detector station (top) and positron energy (bottom)

phases of the horizontal CBO modulation, which lowers sensitivity of ω_a to CBO by a factor of ~ 3.

In the third analysis the data are fitted in 0.2 GeV energy intervals in the range 1.4–3.2 GeV, thus providing important test for ω_a stability versus positron energy. To use positrons with energies less than 2 GeV (which corresponds to the doubled hardware threshold), the standard pileup subtraction procedure was modified. Unlike for the other analyses, here the data are fitted for each of 22 detectors separately (resulting in 198 independent fits) and hence this method does not use some advantages given by symmetric configuration of detectors layout. For this reason, the CBO modulation of phase ϕ_a, neglected in other analyses, is included in set of fit parameters in this one.

In the fourth analyses the ratio $r(t) \sim A \sin(\omega_a t + \phi_a)$, introduced in (6) of [5], was constructed and fitted. This ratio is largely insensitive to changes of observed positron rate on time scales greater than $T_a = 2\pi/\omega_a = 4.37\,\mu s$. Partially by this reason, the effects of asymmetry and phase modulation are not explicitly included in the fitting function.

For all analyses, ω_a was found to be stable (within statistical fluctuations) against the fit start time and run period and, for the third analysis, against the detector number and energy interval, as indicated in Fig. 7. The results are consistent with each other within 0.4 ppm whereas statistical variation of 0.5 ppm is expected due to slightly different data selection and treatment. The overall correction $+0.76 \pm 0.03$ ppm has to be applied due to vertical oscillation of muons and residual effect of electric field.

The combined result of ω_a analyses is $\omega_a/(2\pi) = 229\,074.11 \pm 0.14 \pm 0.07$ Hz, in which the first error is statistical and the second one is systematic. After the ω_a and ω_p were analyzed, separately and independently, a_μ was calculated according to (5) :

$$a_{\mu^+} = \frac{\omega_a/\omega_p}{\mu_\mu/\mu_p - \omega_a/\omega_p} = 11\ 659\ 204(7)(5) \times 10^{-10} \quad (0.7\ \text{ppm})\,. \tag{7}$$

This new result is in good agreement with our previous measurements [3–5]. Weighted average of all of these results gives our final result for a_{μ^+}:

$$a_{\mu^+}(\text{BNL}) = 11\ 659\ 203(8) \times 10^{-10} \quad (0.7\ \text{ppm}) \tag{8}$$

which is, essentially, the new world average. This value, in turn, is in good agreement with the latest measurement at CERN [2]:

$$a_{\mu^+}(\text{BNL}) - a_{\mu^+}(\text{CERN}) = 103(110) \times 10^{-10} \quad (+0.9\ \sigma\ \text{deviation}) \tag{9}$$

The CPT violation probing $(a_{\mu^+} - a_{\mu^-})$ difference has been changed from

$$a_{\mu^+}(\text{CERN}) - a_{\mu^-}(\text{CERN}) = -260(163) \times 10^{-10} \quad (-1.6\ \sigma\ \text{deviation}) \tag{10}$$

to

$$a_{\mu^+}(\text{BNL}) - a_{\mu^-}(\text{CERN}) = -157(120) \times 10^{-10} \quad (-1.3\ \sigma\ \text{deviation})\,. \tag{11}$$

The deviation of some 1.5 σ withstood over the last three decades. That makes our upcoming results on a_{μ^-} even more intriguing.

Current status of experimental value of a_μ is illustrated in Fig. 8.

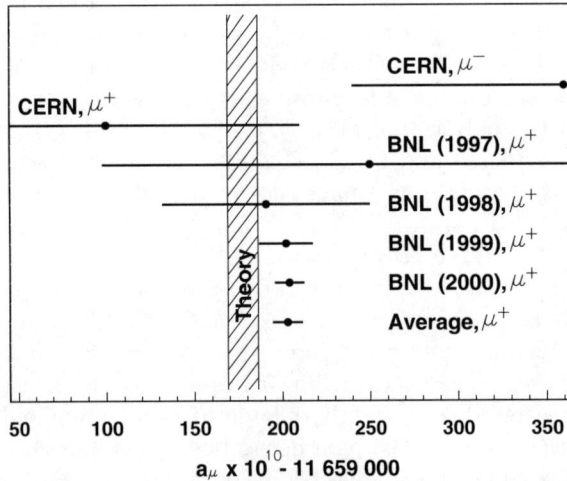

Fig. 8. Current status of experimental and theoretical values for $a_\mu = \frac{1}{2}(g-2)_\mu$. Results for CERN experiment and BNL 1997, 1998, 1999 and 2000 data taking runs are published in [2], [3], [4], [5] and [6], respectively

5 Standard Model Prediction for a_μ

Standard Model prediction for a_μ can be expressed as a sum of QED, hadron and weak interactions contributions:

$$a_\mu(\text{SM}) = a_\mu(\text{QED}) + a_\mu(\text{had}) + a_\mu(\text{weak}) \tag{12}$$

in which $a_\mu(\text{QED}) = 11\,658\,470.57(0.29) \times 10^{-10}$ [12] and $a_\mu(\text{weak}) = 15.1(0.4) \times 10^{-10}$ [13]. In turn, $a_\mu(\text{had})$ is conventionally represented as $a_\mu(\text{had}) = a_\mu(\text{had}, 1) + a_\mu(\text{had}, 2) + a_\mu(\text{had}, \text{lbl})$, where $a_\mu(\text{had}, 2) = -10.1(0.6) \times 10^{-10}$ [14] and $a_\mu(\text{had}, \text{lbl}) = 8.6(3.2) \times 10^{-10}$ [15] are of the higher order terms, the latter of them is hadronic light-by-light scattering contribution. Recent correction of sign of $a_\mu(\text{had}, \text{lbl})$, initiated in part by 2.6 sigma disagreement of our previous measurement value for a_μ [5] with standard model prediction $a_\mu(\text{SM})$ (at that time) [16], was a major improvement of $a_\mu(\text{SM})$.

The lower order hadronic contribution, $a_\mu(\text{had}, 1)$, is currently the main source of error for $a_\mu(\text{SM})$. Strong QCD coupling at low energy precludes full QCD calculation, but $a_\mu(\text{had}, 1)$ can be found from experimental measurement of $e^+e^- \to hadrons$ cross section and the dispersion integration

$$a_\mu(\text{had}, 1) = \frac{1}{4\pi^3} \int_{4m_\pi^2}^{\infty} ds\, \sigma_{e^+e^- \to \text{hadrons}}(s)\, K(s) \tag{13}$$

with known kernel function $K(s)$.

Complimentary to the direct e^+e^- data, the isovector cross sections of $e^+e^- \to hadrons$ (the most important being $e^+e^- \to \pi^+\pi^-$, $\pi^+\pi^-\pi^+\pi^-$ and $\pi^+\pi^-\pi^\circ\pi^\circ$) can be related, with assumption of isospin invariance, to the corresponding τ decay modes through the conserved vector current. Thus for the energy region $s < m_\tau^2$, τ decay data can be used alternatively for the $a_\mu(\text{had}, 1)$ evaluation. Correction of about $(-10.0 \pm 2.5) \times 10^{-10}$ for isospin violation has to be applied. Using e^+e^- and τ decay data available by now, Davier, Eidelman, Höcker and Zhang [17] found:

$$a_\mu(had, 1) = 684.7(7.0) \times 10^{-10} \qquad \text{(from direct } e^+e^- \text{ data)} \qquad \text{and} \tag{14}$$
$$a_\mu(had, 1) = 701.9(6.2) \times 10^{-10} \qquad \text{(from } \tau \text{ decay data)} \tag{15}$$

with statistically independent parts $a_\mu(had, 1)_{2\pi, 4\pi}$ for the e^+e^- and τ decay data being about three standard deviations apart from each other.

With the results for $a_\mu(had, 1)$ given in (14), (15), the total values for $a_\mu(\text{SM})$ are

$$a_\mu(\text{SM}) = 11\,659\,169.1(7.8) \times 10^{-10} \qquad \text{and} \tag{16}$$
$$a_\mu(\text{SM}) = 11\,659\,186.3(7.1) \times 10^{-10}, \tag{17}$$

respectively. The deviations of these values from the experimental result in (8) are 3.0 (e^+e^-) and 1.6 (τ) standard deviations. The gap between theoretical values in (16), (17) is shown in Fig. 8 by the shadow band. It indicates current uncertainty in $a_\mu(\text{SM})$ (other recent evaluations [18] fall into that gap). Obviously, it's extremely important and urgent to clarify the present situation around $a_\mu(had, 1)$ and thus refine the $a_\mu(\text{SM})$ value.

6 Outlook

In addition to our μ^+ data collected in 1997-2000 runs, in 2001 we have collected, and currently are analizing, about 3 billion of electrons from μ^- decays. Field focusing indexes n=0.122 and n=0.142 were used in 2001 in order to reduce effect of the coherent betatron oscillations and thus to lower systematic error. To provide measurement of a_{μ^-} at the same level of accuracy as for a_{μ^+}, we plan further data taking to obtain additional 6 billion decay electrons. Measurement of a_{μ^-} will provide a sensitive test of CPT conservation and also improved value of a_μ.

Further progress may be achieved in upgraded and/or next generation (g–2) experiments, those possibilities currently are being actively discussed within the collaboration.

References

1. R.S. Van Dyck, P. B. Schwinberg and H. G. Dehmelt: Phys. Rev. Lett. **59**, 26 (1987).
2. J. Bailey et al.: Nucl. Phys. B**150**, 1 (1979).
3. R.M. Carey et al.: Phys. Rev. Lett. **82**, 1632 (1999).
4. H.N. Brown et al.: Phys. Rev. D**62**, 091101 (2000).
5. H.N. Brown et al.: Phys. Rev. Lett. **86**, 2227 (2001).
6. G.W. Bennett et al.: Phys. Rev. Lett. **89**, 101804 (2002).
7. W. Liu et al.: Phys. Rev. Lett. **82**, 711 (1999).
8. S. Sedykh et al.: Nucl. Instr. Meth. A**455**, 346 (2000).
9. G.T. Danby et al.: Nucl. Instr. Meth. A**457**, 151 (2001).
10. P. Prigl et al.: Nucl. Instr. Meth. A**374**, 118 (1996).
11. X. Fei et al.: Nucl. Instr. Meth. A**394**, 349 (1997).
12. V.W. Hughes and T. Kinoshita: Rev. Mod. Phys. **71**, S133 (1999); A. Czarnecki and W. Marciano: Nucl. Phys. B (Proc. Suppl.) **76**, 245 (1999).
13. E.A. Kuraev, T.V. Kukhto, A. Schiller and Z.K. Silagadze: Nucl. Phys. B**371**, 567 (1992); A. Czarnecki, B. Krause and W. Marciano: Phys. Rev. Lett. **76**, 3267 (1996); G. Degrassi and G.F. Giudice: Phys. Rev. D**58**, 53007 (1998); A. Czarnecki and W. Marciano: Phys. Rev. D**64**, 013014 (2001).
14. B. Krause: Phys. Lett. B**390**, 392 (1997).
15. M. Knecht, A. Nyffeler, M. Perrottet and E. de Rafael: Phys. Rev. Lett **88**, 071802 (2002); M. Knecht and A. Nyffeler: Phys. Rev. D**65**, 073034 (2002); M. Hayakawa and T. Kinoshita (2001), hep-ph/0112102; J. Bijnens, E. Pallante and J. Prades: Nucl. Phys. B**626**, 410 (2002); I. Blokland, A. Czarnecki and K. Melnikov: Phys. Rev. Lett. **88**, 071803 (2002).
16. R. Alemany, M. Davier and A. Höcker: Eur. Phys. J. **C2**, 123 (1998); M. Davier and A. Höcker, Phys. Lett B**419**, 419 (1998); M. Davier and A. Höcker, Phys. Lett B**435**, 427 (1998).
17. M. Davier, S. Eidelman, A. Höcker and Z. Zhang, Eur. Phys. J. **C27**, 497 (2003).
18. S. Narison, Phys. Lett B **513**, 53 (2001), Erratum-ibid. B**526**, 414 (2002); J. F. de Trocóniz and F. J. Ynduráin: Phys. Rev. D**65**, 093001 (2002).

Part V

Precision Measurements and Fundamental
Constants

Single Ion Mass Spectrometry at 100 ppt and Beyond

S. Rainville, J.K. Thompson, and D.E. Pritchard

Research Laboratory of Electronics, Department of Physics, Massachusetts Institute of Technology, Cambridge, MA 02139, U.S.A.

Abstract. Using a Penning trap single ion mass spectrometer, our group has measured the atomic masses of 14 isotopes with a fractional accuracy of about 10^{-10}. The masses were extracted from 28 cyclotron frequency ratios of two ions altenately confined in our trap. The precision on these measurements was limited by the temporal fluctuations of our magnetic field during the 5–10 minutes required to switch from one ion to the other. By trapping two different ions in the same Penning trap at the same time, we can now simultaneously measure their two cyclotron frequencies and extract the ratio with a precision of about 10^{-11} in only a few hours. We have developed novel techniques to measure and control the motion of the two ions in the trap and we are currently using these tools to carefully investigate the important question of systematic errors in those measurements.

1 Overview

Accuracy in mass spectrometry has been advanced over two orders of magnitude by the use of resonance techniques to compare the cyclotron frequencies of single trapped ions. This paper provides an overview of the MIT Penning trap apparatus, techniques and measurements. We begin by describing the various interesting applications of our mass measurements and the wide-ranging impact they have on both fundamental physics and metrology. In the same section, we also describe further scientific applications that an improved accuracy would open. This serves as a motivation for our most current work (described in Sect. 4) to increase our precision by about an order of magnitude.

Before describing the latest results, we give in Sect. 3 an overview of our apparatus and methods, with special emphasis on the techniques which we have developed for making measurements with accuracy around 10^{-10}. In those measurements, we alternately trapped two different ions (one at the time) and compared their cyclotron frequencies to obtain their mass ratio. The main limitation of this method was the fact that our stable magnetic field would typically fluctuate by several parts in 10^{10} during the 5–10 minutes required to switch from one ion to the other. In order to eliminate this problem, we now confine both ions simultaneously in our Penning trap. In Sect. 4, we describe the various techniques that have allowed us to load a pair in the trap and demonstrate a significant gain in precision from simultaneously measuring both their cyclotron frequencies. New tools to measure and control the motion of the ions are also presented. Those tools are invaluable in our current investigation of the important question

of systematic errors. Unfortunately, because this work is ongoing at the time
of this publication, we cannot report a new mass ratio measurement, but this
new technique shows promise to expand the precision of mass spectrometry an
order of magnitude beyond the current state-of-the-art. Finally, we discuss in
Sect. 5 two other techniques that will address the next source of random error in
our measurements: cyclotron amplitude fluctuations. Both techniques (squeezing
and electronic refrigeration) have already been demonstrated by our group.

It should be noted that in addition to our work in ultra-high precision mass
spectrometry, R. Van Dyck's group at the University of Washington has performed measurements of 7 atomic species and their results for the same ions
agree satisfactorily with our masses [1–3].

2   Scientific Applications

Of the three basic physical quantities – mass, length, and time – mass is currently
measurable with the least accuracy. This is unfortunate because mass uncertainties are often the limiting factor in precision experiments and metrology. Also,
accurate mass differences between initial and final states directly determine the
energy available for a variety of interesting physical and chemical processes (e.g.
emission of a gamma ray, neutrino, neutron, or electron, and chemical reactions).

To date we have measured a total of 14 neutral masses, ranging from the
masses of the proton and neutron to the mass of ^{133}Cs, all with accuracies near
or below 10^{-10} – one to three orders of magnitude better than the previously accepted values [4,5]. Our mass measurements have wide-ranging impact on both
fundamental physics and metrology. The masses of hydrogen and of the neutron are considered fundamental constants [6]. The neutron capture processes
^{12}C(n,γ) and ^{14}N(n,γ) emit gamma rays used as calibration lines in the 2–10
MeV range of the gamma spectrum; our measurement of ^{15}N-^{14}N revealed an
80 eV error (8 times the quoted error) in the most widely used standard and
lowered its error to 1 eV [4]. Our ^{20}Ne measurement resolved a huge discrepancy
(reflected in the old error) involving determination of atomic masses from the
energy of nuclear decay products. Also, our result for the mass of ^{28}Si is necessary for one scheme to replace the artifact kilogram (the only non-physics based
metrological standard) by defining Avogadro's number.

Finally our most recent mass measurements of the four alkali atoms ^{133}Cs,
^{87}Rb, ^{85}Rb and ^{23}Na have opened a new route to the fine structure constant
α [5]. Indeed, a route to α that appears likely to yield a value at the ppb level
is obtained by expressing the fine structure constant in terms of experimentally
measurable quantities as follows (in SI units):

$$\alpha^2 = 2cR_\infty \frac{f_{rec}}{f_{D1}^2} \frac{M_{Cs}}{M_e}.$$

(1)

The Rydberg constant R_∞ has been measured to an accuracy of 0.008 ppb [7],
the frequency of the cesium D1-line f_{D1} was measured by Hänsch's group at the
Max-Planck-Institut in Garching to 0.12 ppb [8], and the mass of the electron

in atomic mass units M_e was recently obtained to 0.8 ppb from a measurement of the g factor of the bound electron in $^{12}C^{5+}$ [9] (in reasonable agreement with the previous value at 2 ppb from VanDyck's group [10]). The mass of ^{133}Cs in atomic mass units M_{Cs} was previously known to 23 ppb and we measured it to 0.2 ppb. Finally, the recoil frequency shift f_{rec} of a Cs atom absorbing photons of laser light at the D_1 line is being measured in Chu's group at Stanford University. There has been recent reports of a value of f_{rec} at or below 10 ppb but it has not been published yet. Combining these results in (1) will lead to a new determination of α with a precision of about 5 ppb. This is similar to the precision of the current best measurement of alpha from the $g - 2$ factor of the electron combined with QED calculations (4 ppb) and can therefore be regarded as a check of QED at an unprecedented level of precision. In addition to resting on such simple and solid physical foundations, this route has the advantage that it can be exploited with many different atomic systems. An experiment is already under way at the Laboratoire Kastler Brossel (ENS, France) to measure the atomic recoil frequency shift in Rb [11] and a new type of interferometer has been demonstrated at MIT (USA) to measure the same quantity in Na [12].

Another interesting application for precise mass measurements involves measuring the mass difference between two atoms related by a neutron capture process, like ^{15}N and ^{14}N. Kessler and collaborators at NIST have precisely measured the wavelengths of the γ-rays emitted in a few neutron capture processes [13]. By comparing the energy of the γ-rays to the mass difference between the initial and final states, one can look for a violation of special relativity. The basic idea is to write

$$\Delta m c_m^2 = \frac{h c_{em}}{\lambda} \qquad (2)$$

in which a photon of wavelength λ is emitted in a process where a mass Δm is converted into electromagnetic radiation. The quantities c_m and c_{em} are respectively defined as the limiting velocity of a massive particle and the velocity of propagation of an electromagnetic wave in vacuum. According to the special theory of relativity, these two quantities are the same, i.e., $c_m = c_{em}$. Independent measurements of λ and Δm could ultimately place limits on the quantity $(1 - c_m/c_{em})$ at the level of $(1 \text{ to } 2) \times 10^{-7}$. This would improve the current limit (from the Compton wavelength of the electron and the von Klitzing constant) by about two orders of magnitude [14]. Unlike other tests of special relativity (Michelson-Morley, Kennedy-Thorndike, etc.), this limit does not depend on assumptions concerning the motion of the laboratory with respect to a preferred reference frame. In order to determine Δm with the same precision as the one reached by the NIST group on λ (few parts in 10^7), we need to be able to make mass comparisons with an accuracy of a few parts in 10^{11}.

Mass being such a fundamental quantity of matter (and one of the three basic physical quantities), it is inevitable that measuring it more precisely will open new possibilities in metrology and challenge our understanding of nature. More specifically, in addition to the application mentioned above, new mass measurements with a precision of 10^{-11} could provide new metrological benchmarks and help determine the mass of the electron neutrino (with the $^3H-^3He$ mass difference) [15,16]. If we could reach our ultimate goal of 10^{-12}, we would have a generally useful technique to directly measure excitation and chemical binding energies of atomic and molecular ions by weighing the associated small decrease in mass, $\Delta E = \Delta mc^2$. A novel technique we have recently developed that shows promise towards achieving those goals will be discussed in Sect. 4.

3 Experimental Techniques

Our atomic masses are determined by comparing the cyclotron frequencies $\omega_c = qB/m$ of single atomic or molecular ions. The ions are held in a Penning trap which consists of a strong uniform magnetic field (8.5 T) and a weak dc quadrupole electric field to confine the ions along the direction of the magnetic field. The electric field is generated by a set of hyperbolic electrodes shown on Fig. 2. Another set of electrodes, called guard rings, are located on the hyperbolic assymptotes and are adjusted to approximately half the voltage on the ring electrode in order to minimize the lowest order non-quadrupole electric field component (C_4). The electrode surfaces are coated with graphite (Aerodag) to minimize charge patches. Together with the magnetic field, they form what is called an orthogonally compensated hyperbolic Penning trap with characteristic size $d = 0.549\,\mathrm{cm}$. At rf frequencies, the guard rings are split in order to provide dipole drives and quadrupole mode couplings for the radial modes (see Sect. 3.2). Figure 1 shows the location of the Penning trap relative to the rest of our apparatus. Trapping the ion allows the long observation time necessary for high precision. Using a single ion is crucial for high accuracy since this avoids the frequency perturbations caused by the Coulomb interaction between multiple ions.

The combination of magnetic and electric fields in our Penning trap results in three normal modes of motion: trap cyclotron, axial, and magnetron, with frequencies $\omega_c'/2\pi \approx 5\,\mathrm{MHz} \gg \omega_z/2\pi \approx 0.2\,\mathrm{MHz} \gg \omega_m/2\pi \approx 0.005\,\mathrm{MHz}$, respectively. The free-space cyclotron frequency ω_c is recovered from the quadrature sum of the three normal mode frequencies (invariant with respect to trap tilts and ellipticity [17]):

$$\omega_c = \frac{qB}{m} = \sqrt{\omega_{c'}^2 + \omega_z^2 + \omega_m^2} \,. \tag{3}$$

We produce ions by ionizing neutral gas in our trap. From a room temperature gas-handling manifold we inject a small amount of neutral gas at the top of our apparatus (Fig. 1) and it diffuses down into the trap through a small hole in the upper endcap. From a field emission tip at the bottom of the trap (shown on Fig. 2), we generate a very thin electron beam (sub-μm diameter) which then ionizes atoms or molecules inside the trap. Since the electron beam is parallel and close to the trap axis, the ions are created with a small magnetron radius ($\leq 100\,\mu\mathrm{m}$). We test for the presence of ions by applying a short drive pulse on the lower end cap and looking for the ions' signal in our detector (see Sect. 3.1). We determine the number of (identical) ions produced by measuring the damping time of the ion signal. If more than two ions are present, we normally invert

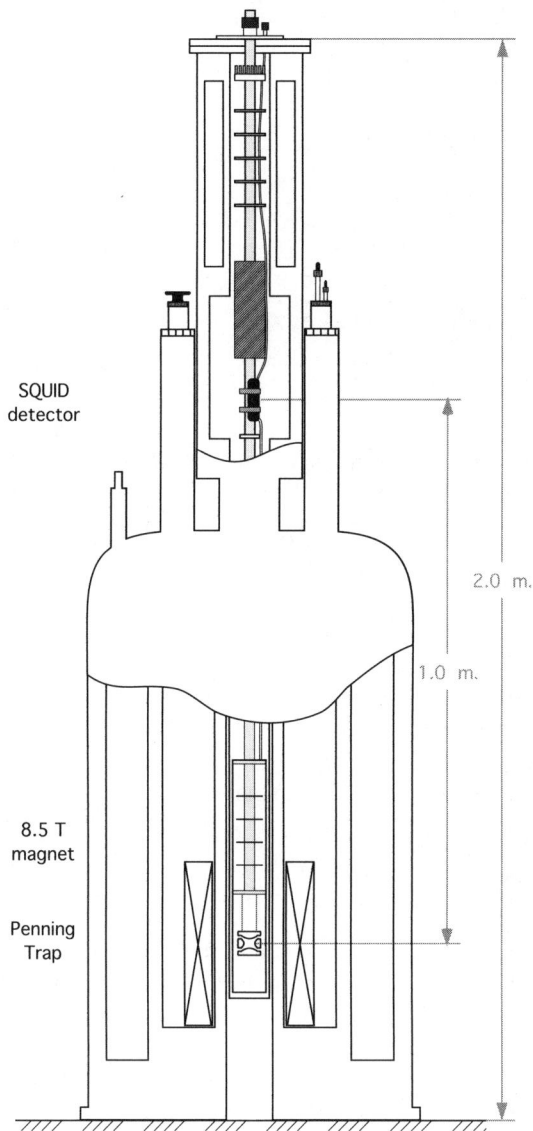

Fig. 1. Schematic of the ion mass spectrometer at MIT. The superconducting magnet produces a stable 8.5 T magnetic field. The image current induced in the endcap by the ion's axial motion is detected using a dc SQUID. The trap, the magnet and the SQUID are at liquid helium temperature (4 K)

the trap and try again. We gradually reduce the amount of gas used, the electron beam current and the time we leave it on until we make, on average, a single ion of the kind we want. Since this ion making procedure is not selective, it sometimes produces unwanted ions with different masses. For example, if we use N_2 gas to make one N_2^+ ion, we might also make N^+ ion(s). We eliminate these so called "fragments" by selectively exciting their axial motion (since they have a different mass/charge ratio, their axial frequency is different) and then bringing the equilibrium position of the ion cloud very near the lower endcap (by applying

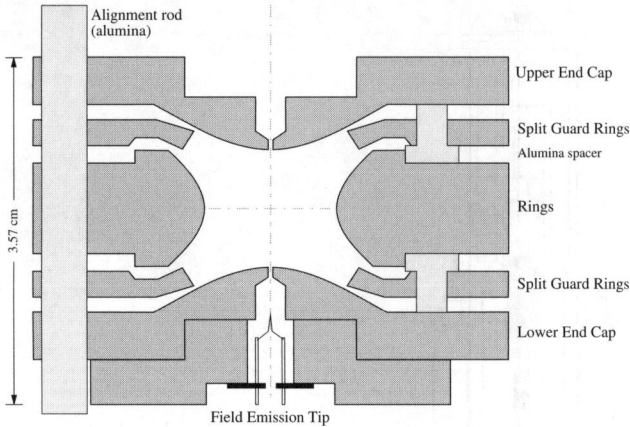

Fig. 2. Cross section of our orthogonally compensated hyperbolic Penning Trap. The copper electrodes are hyperbolae of rotation and form the equipotentials of a weak quadrupole electric field. Guard ring electrodes located on the hyperbolic assymptotes are adjusted to minimize the lowest order non-quadrupole electric field component. The electrode surfaces are covered with a thin layer of graphite (Aerodag) to minimize charge patches

a dc voltage to it). The highly excited ions then neutralize by striking the encap and we are left with only the desired ion in the trap. If 2 or 3 identical ions remain in the trap, bringing them progressively closer and closer to the endcap has a good chance of thining the cloud down to only a single one.

3.1   SQUID Detector

We have developed ultrasensitive superconducting electronics to detect the minis-cule currents ($\leq 10^{-14}$ amperes) that a single ion's axial motion induces in the trap electrodes. The detector consists of a dc SQUID coupled to a hand wound niobium superconducting resonant transformer ($Q \approx 45\,000$) connected across the endcaps of the Penning trap [18]. Our detection noise is currently domi-nated by the 4 K Johnson noise present in the resonant transformer – a fact we have exploited as discussed in Sect. 5.2. Energy loss in the resonant transformer damps the axial motion on a time scale of typically 1 second (at $m/q \approx 30$), quickly bringing the axial motion to thermodynamic equilibrium at 4 K. Since the ion signal is concentrated in a narrow frequency band, we can easily detect it against the broad Johnson noise with a signal-to-noise ratio of about 10.

Due to the superconducting nature of our detector, both the SQUID and the transformer have to be located about 1 m away from the trap in a region of relatively low the magnetic field (see Fig. 1). They are both encased in sepa-rate superconducting niobium boxes wrapped with lead. When the apparatus is cooled down, external bucking coils zero the magnetic field at the location of the detector. Once the niobium boxes are superconducting, they keep magnetic flux out and allow the operation of the detector without any current in the bucking coils.

3.2   Mode Coupling and π-Pulses

The axial oscillation frequency of any ion can be tuned into resonance with the fixed detector frequency by changing the dc trapping voltage. To be able to measure the cyclotron frequency using only our axial mode detector, we use a resonant rf quadrupole electric field which couples the cyclotron and axial modes [19]. This coupling causes the two modes to cyclically and phase coherently exchange their classical actions (amplitude squared times frequency). In analogy to the Rabi problem, a π-pulse can be created by applying the coupling just long enough to cause the coupled modes to exactly exchange their actions. The same rf quadrupole field is also used to cool the cyclotron mode by coupling it continuously to the damped axial mode. By using a different rf frequency, the exact same technique can be used to measure and cool the magnetron mode.

3.3   Pulse and Phase Technique

We have developed the Pulse aNd Phase (abbreviated PNP) method to achieve a relative uncertainty of 10^{-10} on cyclotron frequency measurements in less than 1 minute [20]. A PNP measurement starts by cooling the trap cyclotron mode via coupling to the damped axial mode (see Sect. 3.2). The trap cyclotron motion is then driven to a reproducible amplitude and phase at $t = 0$ and then allowed to accumulate phase for some time T, after which a π-pulse is applied. The phase of the axial signal immediately after the π-pulse is then measured with rms uncertainty of order 10 degrees. Because of the phase coherent nature of the coupling, this determines the cyclotron phase with the same uncertainty. The trap cyclotron frequency is determined by measuring the accumulated phase versus evolution time T with the shorter times allowing the measured phase (which is modulo 360 degrees) to be properly unwrapped. Since we can typically measure the phase within 10 degrees, a cyclotron phase evolution time of about 1 minute leads to a determination of the cyclotron frequency with a precision of 10^{-10}.

The PNP method has the advantage of leaving the ion's motion completely unperturbed during the cyclotron phase evolution [19]. It is also particularly suited to measure mass doublets – pairs of species such as CD_4^+ and Ne^+ that have the same total atomic number. Good mass doublets typically have relative mass difference of less than 10^{-3}, making these comparisons insensitive to many systematic instrumental effects.

3.4   Separate Oscillatory Field Technique

To compare an ion to ^{12}C, it is crucial to determine the masses of ^1H and D (^2H) so that they can be combined with ^{12}C to form doublet comparison molecules (for instance O^+/CH_4^+ and Ne^+/CD_4^+) since comparing near equal masses reduces the size of many experimental systematic errors. Therefore, it is crucial to determine the masses of ^{1}H and D. However, there are very few routes for doing this with doublet comparisons and even fewer direct routes involving a single mass doublet comparison.

To illustrate why it is difficult to find a series of doublet mass ratios which yield masses of ^{1}H and D, consider the set of comparisons (i.e. mass ratios)

$$\frac{N^+}{CH_2^+} , \quad \frac{O^+}{CH_4^+} , \quad \frac{CO^+}{N_2^+}$$

which would seem to determine the three unknown atomic masses H, N and O (i.e. relative to C). A doublet mass ratio is so close to 1 that it should be thought of as determining a mass difference. For example, if $R \equiv N^+/CH_2^+$ then

$$N^+ - CH_2^+ \approx (R-1)(CH_2^+)' = \Delta M < 0.001 \times (CH_2^+)'$$

where the mass $(CH_2^+)'$ is known from other experiments with several orders of magnitude less accuracy. From this perspective and after correcting for ionization and molecular binding energies, the above set of measured cyclotron frequency ratios determine the mass differences

$$N - C - 2H = \Delta M_1 ,$$
$$O - C - 4H = \Delta M_2 ,$$
$$O + C - 2N = \Delta M_3 .$$

Unfortunately, it is clear that these relations yield only 2 linearly independent equations– combining the first two equations yields the third which therefore is a consistency check on the three measurements. Using non-doublet ratios such as CD_4^+/C^+ and CH_4^+/C^+ removes such singularities from the matrix relating neutral atomic masses to measured mass differences.

In the case of a non-doublet comparison, the difference in the trapping voltages needed to detect each ion's axial motion is large enough to cause significant shifts in the ion's equilibrium position due to charge patches on the trap electrodes. The shift in equilibrium position causes a systematic error because of magnetic field inhomogeneities.

We have developed the SOF (separated oscillatory field) technique [21] to allow us to make cyclotron frequency comparisons using the same trapping voltage during the phase evolution time. An SOF sequence is identical to the PNP sequence but with a second drive pulse equal in strength to the first in place of the π-pulse. If the second drive pulse is in (out of) phase with the cyclotron motion, the two drive pulses add (subtract) resulting in a large (small) cyclotron amplitude. The result is that the cyclotron's phase information is encoded in the cyclotron amplitude. The trapping voltages can be adiabatically adjusted for axial detection and a π-pulse then applied. The detected axial amplitude versus phase evolution time T produces a classical Ramsey fringe which oscillates at the difference between the drive and trap cyclotron frequencies.

3.5   Making a Mass Table

A cyclotron frequency ratio of two different ions is determined by a run measuring a cluster of ω_c values for an ion of type A, then for type B, etc. In a typical 4-hour run period (from 1:30-5:30 am when the nearby electrically-powered subway is not running), we can typically record between 5 and 10 alternations of ion type (see Fig. 3).

Since the measured free-space cyclotron frequencies exhibit a common slow drift, we fit a polynomial plus a frequency difference to the combined set of cyclotron frequency measurements for the night. The average polynomial fit order is typically between 3 and 5 and is chosen using the F-test criterion [22] as a guide to avoid removing frequency changes which are not correlated between ion types. The distribution of residuals from the polynomial fits has a Gaussian center with a standard deviation $\sigma_{\text{resid}} \approx 0.25$ ppb and a background ($\approx 2\%$ of the points) of non-Gaussian outliers [4]. We handle these non-Gaussian outliers using a robust statistical method to first properly describe the observed statistical distribution of data points and then to smoothly deweight the nongaussian points [23,4].

In all, we have measured a set of 28 cyclotron frequency ratios during 55 night runs using the techniques described above. In order to convert those ion mass ratios to mass differences of neutral isolated atoms, we account for chemical binding energies and for the mass of the missing electrons and their ionization energies [24,25]. Because those are small corrections and they are known with enough precision, they don't contribute to our final uncertainties. Performing a global least square fit to all these linear equations yields the neutral atomic masses in Table 1. The fit produces a covariance matrix which directly yields the uncertainty in the atomic mass and allows uncertainties to be calculated for quantities involving correlated isotopes. Atomic masses are expressed relative

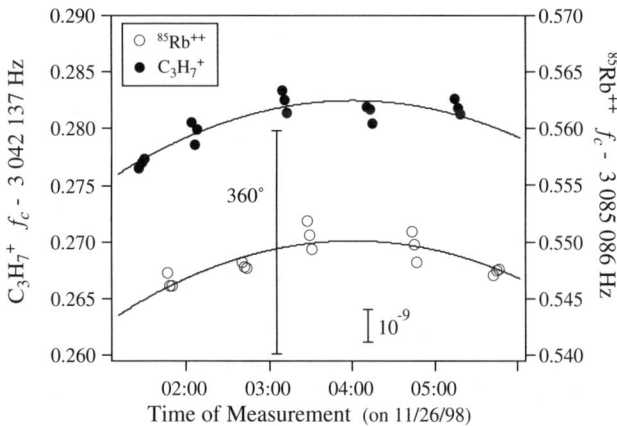

Fig. 3. Typical set of data comparing the cyclotron frequencies of two single ions alternately loaded into the Penning Trap. Magnetic field drifts (fitted here with a polynomial) limit the relative precision on the mass ratio to about 10^{-10}

Table 1. Neutral masses measured at MIT

Species	MIT Mass (u)	ppb	$\frac{\sigma_{1983}}{\sigma_{MIT}}$
^{1}H	1.007 825 031 6 (5)	0.50	24
n	1.008 664 916 4 (8)	0.81	17
^{2}H	2.014 101 777 9 (5)	0.25	48
^{13}C	13.003 354 838 1 (10)	0.08	17
^{14}N	14.003 074 004 0 (12)	0.09	22
^{15}N	15.000 108 897 7 (11)	0.07	36
^{16}O	15.994 914 619 5 (21)	0.13	24
^{20}Ne	19.992 440 175 4 (23)	0.12	957
^{23}Na	22.989 769 280 7 (28)	0.12	93
^{28}Si	27.976 926 532 4 (20)	0.07	350
^{40}Ar	39.962 383 122 (33)	0.08	424
^{85}Rb	84.911 789 732 (14)	0.16	193
^{87}Rb	86.909 180 520 (15)	0.17	187
^{133}Cs	132.905 451 931 (27)	0.20	111

to ^{12}C, which is defined to have a mass of exactly 12 atomic mass units (u). In Table 1, the error in the last digits is in parenthesis. The last two columns give the fractional accuracy of the measurements in ppb (parts in 10^9) and the improvement in accuracy over pre-Penning Trap mass values (from the 1983 atomic mass evaluation [26]). Note that the mass of the neutron is determined from the masses of ^{1}H and ^{2}H combined with measurements of the deuteron binding energy which has recently been improved [27].

The measured ratios were chosen so that at least two completely independent sets of mass ratios enter into the determination of each atomic mass. The overall $\chi_\nu^2 = 0.83$, indicating excellent internal consistency. Other experimental checks on systematic errors include measuring calculable mass to charge ratios (i.e. Ar^{++}/Ar^+) and measuring redundant mass ratios at different mass to charge ratios which would have very different systematic errors (i.e. O^+/CH_4^+, $CO^+/C_2H_4^+$, and $CO_2^+/C_3H_8^+$ all determine the same mass difference C + 4H - O at $m/q = 16$, 28, and 44).

Regarding the important question of systematic errors, our group has been in a somewhat unique situation for precision experiments. All the possible systematic errors were estimated (and some of them experimentally tested) to be well below the level of the random errors introduced by our magnetic field fluctuations. In other words, the errors on our measurements are entirely dominated by statistical noise from the magnetic field. The various self-consistency checks mentioned above confirmed that this is really the case and no unknown systematic errors are lurking at the level of our errors. The only exception to this situation was in the measured ratios involving Cs and Rb (the heaviest elements in our mass table). When repeatedly measuring those mass ratios for several nights, we found variations larger than our estimated error bar for each night ($\chi^2_\nu \approx 5$). Despite extensive research, we never identified the source of those excess night-to-night variations and we increased the error bars of our reported results to account for those fluctuations (see [5] for more details).

4   Simultaneous Measurements

Until the year 2000, we determined mass ratios by alternately creating individual ions of the two species being compared and measuring their cyclotron frequencies separately as described in the previous section. The precision of this technique is limited almost entirely by temporal fluctuations of the magnetic field, which are typically 3 parts in 10^{-10} during the several minutes required to trap a new single ion. We were also restricted to take precision cyclotron frequency measurements only during the period between about 01:00 and 05:30 at night during which Boston's electric subway is not running (it creates random fluctuations of about 4×10^{-7} T in our lab). In the fall of the year 2000, to avoid the effect of magnetic field fluctuations, we decided to make *simultaneous* measurements of the cyclotron frequencies of the two ions being compared.

Simultaneously comparing the cyclotron frequencies of two different ions in the same trap offers the best protection against magnetic field fluctuations and field gradients, but introduces new complications: ion-ion perturbations and systematic shifts due to spatial field inhomogeneities. In previous work studying the classical, two-body problem of two ions in a single Penning trap [28], we found that if we keep the distance between the ions large enough, the several kHz difference between the two ions' cyclotron frequencies keeps the two cyclotron modes independent from each other. Similarly, the axial frequencies of the two ions are different enough ($\Delta f_z \approx 50$ Hz) to keep the two axial modes uncoupled.

In contrast, since ω_m is to first order independent of mass, the Coulomb interaction between the ions couples the nearly frequency-degenerate magnetron modes into two new collective magnetron modes: a center-of-mass (COM) mode and a difference mode [28]. The COM mode corresponds to the center-of-mass of the ions moving at the average magnetron frequency (~ 5 kHz) about the center of the trap. The difference mode corresponds to an $E \times B$ drift of the ions about the center-of-mass due to the Coulomb interaction between them. The frequency of the difference mode is ~ 50 mHz higher than that of the COM mode. In [28], we showed that in a perfect trap the ion-ion separation distance ρ_s, i.e., the amplitude of the difference mode is constant in time, owing to conservation of energy and canonical angular momentum.

(a)   (b)

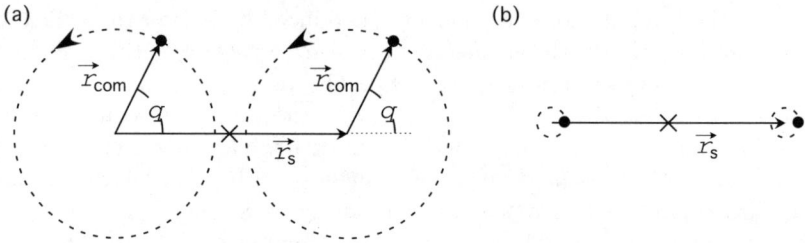

Fig. 4. Two ion magnetron mode dynamics. The magnetic field is pointing out of the plane of the figure and the center of the trap is indicated by a cross. At any time, the position of both ions can be described by the center-of-mass vector ($\boldsymbol{\rho}_{\text{com}}$) and the separation vector ($\boldsymbol{\rho}_{\text{s}}$). Both vectors rotate clockwise at nearly the same frequency (around 5 kHz), but the separation vector rotates about 50 mHz faster due to the Coulomb interaction between the ions (for $|\boldsymbol{\rho}_{\text{s}}| \approx 1\,\text{mm}$). So in a frame where the separation vector is stationary, the ions trace out counter-clockwise tandem circles centered on opposite sides of the center of the trap. (**a**) If $|\boldsymbol{\rho}_{\text{com}}| \approx |\boldsymbol{\rho}_{\text{s}}|/2$ each ion moves in and out of the center of the trap every 20 s. (**b**) If the center-of-mass mode is cooled, the ions are "parked" on nearly the same magnetron orbit. This configuration is preferable for taking cyclotron frequency data

The ideal configuration for a precise comparison of the two cyclotron frequencies is to have the two ions to go around the trap center on a common magnetron orbit of radius 500 μm, always 1 mm apart from each other. In other words, we want a magnetron difference mode amplitude of $\rho_{\text{s}} \sim 1\,\text{mm}$ and a magnetron COM mode amplitude as small as possible as shown in Fig. 4b. This configuration insures that both ions sample the same average magnetic and electrostatic fields while minimizing the ion-ion perturbations of the cyclotron frequencies. This is very important to avoid systematic errors since we know that our trapping electrostatic and magnetic fields are not perfectly homogeneous. For example, the cyclotron frequency of one ion is shifted by about 8 parts in 10^{10} between the center of the trap and a magnetron radius of 500 μm because of magnetic field inhomogeneities. In order to achieve this ideal configuration, we have developed novel techniques to precisely measure and control the individual motion of two different single ions in our Penning trap.

4.1   Two–Ion Loading Techniques

To introduce a pair of ions in the trap, we simply make the two ions one after the other using the procedure described in Sect. 3. However, since ρ_{s} is fixed by the separation distance between the ions when their magnetron modes couple, we must avoid making them too close to each other. Recall from Sect. 3 that our ions are created near the center of the trap, in a magnetron orbit of typically ≤ 100 μm radius. So the basic sequence for loading a pair is the following: we produce a single ion of the first member of our pair, say $^{13}\text{C}_2\text{H}_2^+$. We choose to make the more difficult ion to isolate first since we can then apply our cleaning techniques to remove fragments (such as $^{13}\text{C}_2\text{H}^+$, $^{13}\text{C}_2^+$, etc.). After cooling all of the first ion's three modes of motion, we drive its magnetron radius to ~ 1 mm and then make the second ion (N_2^+) near the trap center. The ion-ion separation ρ_s is then fixed to ~ 1 mm and the amplitude of the magnetron COM $\rho_{com} \approx \rho_s/2 \approx 500$ μm.

The motion of the two ions in this configuration is determined by the two new magnetron normal modes (COM and difference) and is illustrated in Fig. 4a. The frequency of the COM and difference modes are nearly identical (around 5 kHz) but the difference mode rotates about 50 mHz faster due to the Coulomb interaction between the ions [28]. So in a frame where the separation vector is stationary, the ions trace out counter-clockwise tandem circles centered on opposite sides of the center of the trap. This means that each ion's magnetron radius is oscillating between 0 and 1 mm every 20 seconds. The positive aspect of this "swapping motion" is that it insures that both ions experience the same average magnetic and electric fields. However, in the presence of electrostatic anharmonicities the axial frequency of an ion depends on its radial position in the trap. Since our detector relies on the narrowband nature of the ion's signal (Sect. 3.1), the less stable the axial frequency is, the more difficult it is to observe the axial motion of the ions and extract precise information from it (e.g. phase). In order to stabilize the axial frequency and reach the ideal configuration mentioned in the previous section, we need to cool the magnetron COM mode to the configuration shown in Fig. 4b.

One approach for cooling the COM motion is to apply a magnetron drive pulse with the correct amplitude and phase to drive the magnetron center-of-mass to the center of the trap. To show that we can do this, we first put a single $^{13}C_2H_2^+$ ion in the trap and drive its magnetron motion with two magnetron pulses of equal amplitude separated by a time T (analogous to the SOF technique described in Sect. 3.4). We then minimize the final magnetron amplitude by varying the relative phase between the drive pulses while holding T fixed. At the optimal phase between the drives ϕ_{min}, the ion is pulsed out to a large magnetron radius (typically 1 mm), allowed to go around its magnetron orbit at 5 kHz for T seconds, and then pulsed back to the center of the trap. We can determine ϕ_{min} with uncertainty less than one degree in about 10 min. The remaining magnetron amplitude is typically less than 100 μm. We normally use $T \approx 1$ s but we have been able to observe similar performance with T up to 10 s.

Once we know ϕ_{min}, we have what we need to introduce a pair of ions in our trap and cool their magnetron center-of-mass. We can simply use the same procedure as above with two differences: we make an N_2^+ ion near the trap center during the time T, and we make the amplitude of the second magnetron drive only half of the amplitude of the first one. Here is the sequence in details: 1) We drive $^{13}C_2H_2^+$ to a large 1 mm magnetron orbit. 2) We quickly inject N_2 gas and fire our electron beam to create one N_2^+ ion near the center of the trap. 3) T seconds after the initial magnetron pulse, we apply another magnetron drive with a phase ϕ_{min} relative to the first one and only half the amplitude. The effect of this pulse is to drive the $^{13}C_2H_2^+$ back to a radius of 500 μm, and simultaneously drives the N_2^+ out to a radius of 500 μm on the other side of the trap. The two ions should then be in the "ideal" configuration pictured in Fig. 4b.

By completely automating the ion making process we were able to execute this sequence in less than one second.

The main problem of this method is that since our electron beam is parallel to the trap axis, we often make the N_2^+ with a large axial amplitude (several mm). This greatly increases the effective distance between the two ions such that the two individual magnetron modes do not couple initially. During the several tens of seconds required to damp this axial excitation, the separation distance between the ions is not conserved and the ions are likely to end up much closer to each other than we intended. To avoid this problem we modified the method above to allow axial cooling of the N_2^+ before sending the correcting magnetron drive pulse. Just before making the N_2^+, we make the trap very anharmonic to intentionally break the degeneracy between the magnetron frequencies of the two ions (one in the center of the trap, the other in a 1 mm magnetron orbit). This allows us to cool the axial motion of the N_2^+ without worrying about the magnetron motions swapping amplitudes. However, because the cooling process takes minutes we now need to measure the phase of the $^{13}C_2H_2^+$ in order to choose the phase of our second magnetron drive. To do this, we go back to a harmonic trap and apply a short coupling pulse between the magnetron and axial modes of the $^{13}C_2H_2^+$ and extract its magnetron phase from the signal in our detector. The change in magnetron amplitude from the coupling pulse is insignificant. In principle, we could extend this technique to measure the amplitude and phase of both ions' magnetron motions and send a correction pulse to fine-tune the magnetron orbits of the ions, i.e., zero more precisely the magnetron COM mode amplitude.

Loading a pair of ions in our trap using the techniques above still requires some work, time and patience. We often have to try many times ($\sim$10) before making a pair that we can use. Most often the COM amplitude is still large and the ions are too close to each other so that the axial frequencies of the ions vary a lot very quickly (2–3 Hz in 10 s). However we are rewarded by being able to keep the same pair in the trap and perform measurements on it for many weeks.

4.2   Diagnostic Tools

By simultaneously trapping two different ions in our Penning trap, we introduce two new possible sources of systematic errors on our measurement of the cyclotron frequency ratio: (1) the Coulomb interaction between the ions and (2) the imperfection of the trapping fields away from the center of the trap. In contrast to our previous technique where we altenately trapped single ions (Sect. 3), we expect that these systematic errors will now completely dominate our final error. In order to investigate this important question, it is therefore crucial for us to be able to measure the ion-ion separation, know which part of the trap each ion samples, and precisely characterize our trapping fields. In this section, we will describe various techniques we invented to achieve this.

During the past few years, we have developed a new computer system to control our experimental setup which allows a much higher level of automation. We also accurately measured the relativistic cyclotron frequency shift versus cyclotron radius for a single ion of Ne^{++} and Ne^{+++} and obtained from this an absolute calibration of the amplitude of our ion's motion in the trap with an accuracy of 3% [29]. This allowed us to precisely map the axial and magnetron frequency shifts of one ion in the trap as a function of absolute magnetron, cyclotron and axial amplitudes, from which we gained unprecedented knowledge of our field imperfections (electrostatic and magnetic). Using the conventions of [17] to expand the fields, we have $B_2/B_0 \approx (61 \pm 6) \times 10^{-8}/\text{cm}^2$, $C_6 \approx (53\pm7)\times10^{-4}$ and we can adjust C_4 to the desired value (usually zero) $\pm1\times10^{-5}$ using the guard ring electrodes voltage. We also developed a computer-based feedback system to lock the axial frequency of an ion in the trap to an external frequency reference. This system now allows us to continuously monitor the axial frequency of an ion. Equipped with these new tools, we are now able to exploit the dependence of the axial frequency on magnetron radius in the presence of electrostatic anharmonicities to indirectly observe the radial position of one ion in time. This technique applied to a pair of simultaneously trapped ions allows us to experimentally observe the beat frequency between the strongly coupled magnetron modes (as each ion's magnetron radius oscillates due to a non-zero magnetron COM mode amplitude). Not only has this confirmed our model of the dynamics of two trapped ions, but it also provides us with a sensitive probe of the ion-ion separation distance. Indeed, the magnetron beat frequency scales like the inverse cube of the ion-ion separation distance. We can therefore measure where the ions are with respect to each other with a precision of a few percent and verify that their separation is constant in time at that level. This is an invaluable tool in our exploration the new systematic error on the ratio introduced by the ion-ion Coulomb force. By keeping the ions ~1 mm apart, we expect the cyclotron frequency ratio to be perturbed by less than one part in 10^{11}.

We can also determine the rms magnetron radius of each ion individually by varying the size of the electrostatic anharmonicity and measuring the change in each ion's axial frequency. This is also crucial since we must know that both ions sample the same region of space to prevent magnetic field inhomogeneities from introducing a systematic error.

An unexpected effect of our axial frequency locking system is that it appears to couple our two-ion magnetron normal modes. Depending on the frequency of our cw drive relative to the axial frequency of the ion we are locking, we found that we can transfer angular momentum either from the COM mode into the difference mode or vice versa. This is a very useful tool since it allows us to completely cool the magnetron COM amplitude with the important benefits mentioned above. It also gives us the ability to change the ion-ion separation without having to load a new pair of ions in our trap. Indeed we can drive the COM magnetron mode with a dipole electric field and then use this technique to transfer that COM mode amplitude into the difference mode, thereby moving the ions further apart from each other. To reduce the ion-ion separation distance, we simply use a short axial-magnetron coupling pulse to transfer a little bit of the magnetron motion into the damped axial mode for both ions simultaneously.

4.3   Preliminary Results

Using the techniques described in the previous sections, we have been able to load the trap with two different ions species (e.g. $^{13}C_2H_2^+$ and N_2^+) and simultaneously confine them on nearly the same magnetron orbit in our Penning trap (with radius of about $500\,\mu m$). We can then apply the same techniques we used previously to measure the cyclotron frequency of a single ion in the trap (see Sect. 3), but on both ions *simultaneously*: we drive each ion's cyclotron mode to a radius of about $75\,\mu m$, let it accumulate phase for some time, and then simultaneously transfer each ion's cyclotron motion into its axial mode to read its phase. Since we have been using two ions with very similar masses ($\Delta m/m \approx 4 \times 10^{-4}$) the two axial signals are very close in frequency ($\Delta f_z \approx 50\,Hz$) and they both fall within the bandwidth of our detector. The result of simultaneous cyclotron frequency comparisons is shown in Fig. 5. In these data, the shot-to-shot noise in the ratio of the cyclotron frequencies is $\sim 7 \times 10^{-11}$ after only three minutes of phase evolution – more than a factor of 10 gain in precision compared to our previous method. We have made simultaneous cyclotron frequency comparisons for periods as long 60 hours all under automated computer control and even during the daytime when magnetic field noise from the nearby subway would prevent comparisons of alternate single ions with useful precision. Unfortunately, at the time of this publication we cannot report a measured mass ratio because our study of the systematic errors associated with having both ions in the trap is still ongoing.

When measuring the cyclotron frequency ratio of CO^+/N_2^+, we have repeatedly observed abrupt and very large (~ 1 part in 10^9) jumps between a few

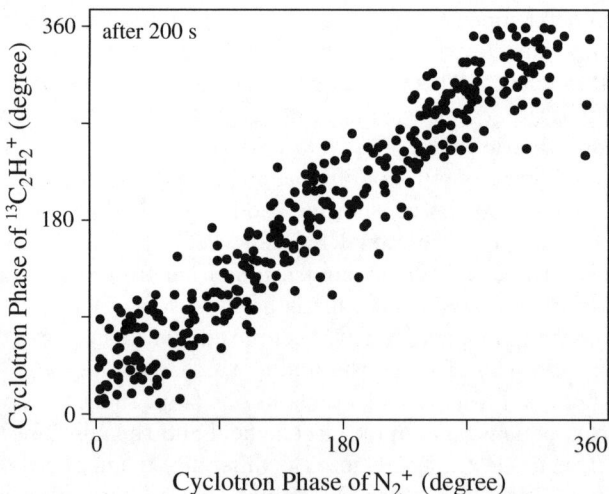

Fig. 5. Preliminary data from two different ions simultaneously confined in the same trap. Each point represents a set of cyclotron phases simultaneously accumulated in 200 s by a $^{13}C_2H_2^+$ and a N_2^+ ion plotted versus each other. The very good correlation indicates that magnetic field fluctuations are not a limitation in this technique

discrete values. No such jumps have been observed in the cyclotron frequency ratio for the experimentally very similar comparison $^{13}C_2H_2^+/N_2^+$. We currently attribute the observed jumps to black-body-induced quantum jumps among the lowest lying rotational levels of the CO^+ molecule. The cyclotron frequency is perturbed because the magnetic field orients the molecular dipole towards or away from the center of cyclotron motion depending on the molecule's rotational state. This is believed to be the first observation of the charge distribution within the molecule modifying its cyclotron frequency and might be used to determine the dipole moments of ionic molecules or for single ion molecular spectroscopy. The details of these results will be published in the next few months.

5 Subthermal Detection

For simultaneous cyclotron frequency measurements, the leading source of random error is the frequency shift associated with thermal variations in the cyclotron radius. This section will describe two techniques which we have already demonstrated to alleviate this problem.

Before every measurement, we cool the ion's motion by coupling it to our detection circuit until it comes into equilibrium with the detector. (Only the axial motion is coupled to the detector, but we cool the two radial modes using the mode coupling field mentioned in Sect. 3.2.) This remaining "4 K" motion of the ion adds vectorially to the displacement from our cyclotron drive pulses and hence prevents us from establishing an exactly reproducible amplitude and phase of motion with each excitation pulse. This effectively adds random noise to the phase we measure from one PNP sequence to the next. Since it is the same Johnson noise that drives the ion's thermal motion and is added to the ion image current to form our detected signal, these two sources of noise both contribute to our measurement error (phase noise). Moreover, the thermal cyclotron amplitude fluctuations cause relativistic mass variations and also combine with field imperfections to introduce random fluctuations of the cyclotron frequency. The cyclotron frequency variations due to special relativity is several parts in 10^{11} for $m/q \sim 20$ and close to a part in 10^{10} for lighter species such as ^{3}He and ^{3}H. After magnetic field fluctuations, this is the dominant source of noise in our alternating measurements technique, and is leading source of random error for simultaneous cyclotron frequency measurements.

5.1 Classical Squeezing

In analogy to squeezed states of light, we have demonstrated a technique in which a parametric drive at $2 \times \omega_z$ produces quadrature squeezing of the axial thermal uncertainty (Fig. 6) [30]. The squeezed thermal distribution can then be swapped into the cyclotron mode using a π-pulse. By properly adjusting the relative phase of the parametric drive and the cyclotron drive, we have demonstrated a factor of 2 reduction in the amplitude fluctuations [30]. We have also proposed two other techniques combining squeezing with selective anharmonicity that should achieve amplitude squeezing by at least a factor of 5 [31].

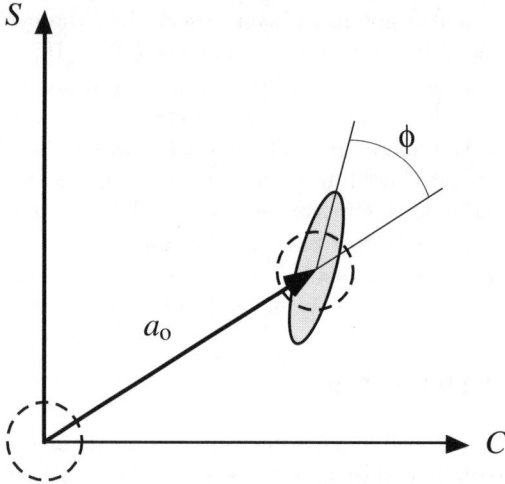

Fig. 6. Squeezing of Thermal Distribution. The finite temperature of the ion's cyclotron mode results in shot-to-shot variation of the cyclotron amplitude and phase after the initial drive pulse (indicated by a_o) of a PNP or SOF sequence. The isotropic thermal uncertainty (*dashed circle*) can be squeezed (*ellipse*) before the drive pulse to reduce either the amplitude ($\phi = 90°$) or the phase ($\phi = 0°$) uncertainty. Amplitude squeezing is more advantageous for mass measurements because amplitude fluctuations lead to shot-to-shot cyclotron frequency fluctuations due to special relativity and magnetic field inhomogeneities

5.2   Electronic Refrigeration

The other approach to address the problem of thermal variations in the cyclotron radius is to cool the detector and ion below the 4 K ambient temperature of the coupling coil and trap environment. This is done with electronic cooling [32]. This technique has the added benefit of greatly improving our signal-to-noise ratio.

The essence of electronic cooling is to measure the thermal noise in our detection transformer (referred to as the coil below), phase shift the signal and then feed it back into the detection circuit to reduces the noise currents to an effective temperature as low as 0.5 K. The key is that our dc SQUID has technical noise much lower than 4 K and can measure precisely the current in the coil in a time shorter than its thermalization time ($Q_0/\omega \sim 30$ ms). This feedback also decreases the apparent quality factor Q of the coil. Figure 7 shows the thermal noise of the coil at different gain settings of the feedback loop. Analyzing these data, we find that the thermal energy in the coil, corresponding to the area under the peak, is reduced below 4 K by the factor Q/Q_0, as expected from the detailed solution of the circuit (assuming a parallel LRC coupling coil where the resistor $R = Q_0\omega_0 L$ has the usual Johnson noise current).

With this electronic cooling technique, the ion's motion should come into equilibrium with the colder detector thereby greatly reducing the problem from amplitude fluctuations described above. Combining this with the squeezing tech-

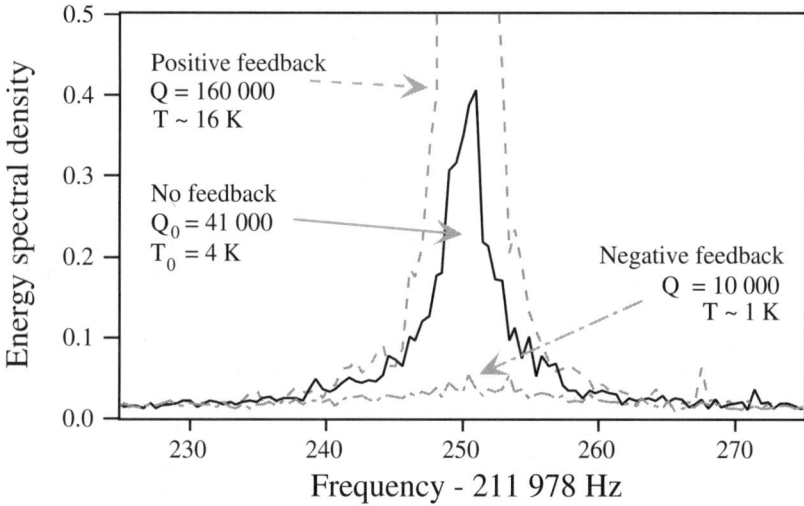

Fig. 7. Thermal profile of the detector coil as a function of the quality factor Q adjusted with the gain of the feedback. The thermal energy in the coil (area under the peak) is proportional to Q/Q_0, where Q_0 is the Q of the detector coil without feedback. This shows that the negative feedback does indeed reduce the thermal fluctuations in the coil

nique described in Sect. 5.1 should in principle reduce the shot-to-shot relativistic fluctuations of the cyclotron frequency ratio to a few parts in 10^{12} for all but the lightest species.

Another effect of the feedback is to reduce the transformer voltage across the trap which is responsible for damping the ion's axial motion. This reduces the bandwidth of our signal, increasing our signal-to-noise ratio (the Johnson noise is a constant current$/\sqrt{\mathrm{Hz}}$). This translates directly into a better ability to estimate the parameters of the axial oscillation of the ion. With this technique, we can now measure the phase of the cyclotron motion of a single ion in the trap with an uncertainty as low as 5 degrees – more than a factor of 2 improvement. Our ability to determine the amplitude of the ion signal has also improved, again by more than a factor of 2, and we can measure the frequency of the axial motion with 4 times better precision. The better phase noise allows us to obtain the same precision on a cyclotron measurement in a shorter time. This will be important in the future since we would have to acquire data for 10 minutes to reach a precision of 10^{-11} with the previous phase noise ($\sim$ 12 degrees), or 100 minutes for 10^{-12} ! We can also use the improved signal-to-noise to reduce the cyclotron amplitude we use, which in turn reduces the frequency shifts due to relativity and field imperfections. Finally, this technique gives us the ability to arbitrarily select the damping time of the ion by changing the gain of the feedback. This opens the door for us to very high precision at small mass-to-charge ratio, (e.g. $^{6,7}\mathrm{Li}$, $^{3}\mathrm{He}$, $^{3}\mathrm{H}$) where we used to suffer from excessively short ion damping times.

6   Conclusion

To date our group has measured a total of 14 neutral masses with fractional accuracies near or below 10^{-10} with wide-ranging impact on both fundamental physics and metrology. This typically represents an improvement of two orders of magnitude in precision over the previously accepted values. This precision was achieved by comparing the cyclotron frequencies of two single atomic or molecular ions alternately confined in a Penning trap. The magnetic field fluctuations limited the precision of a given mass ratio to a few parts in 10^{-10} for a 4 hour data set during the night (when the magnetic field is quiet).

We have now successfully loaded two different single ions in the same Penning trap and demonstrated *simultaneous* measurements of their cyclotron frequencies. This technique completely eliminates the temporal variations of magnetic field as a limitation in our measurements and allows us to attain a shot-to-shot noise in the ratio of $\sim 7 \times 10^{-11}$ after only three minutes of measurement time. This represents more than a factor of 10 gain in precision compared to our previous method. We have developed novel techniques to measure and control all three normal modes of motion of each ion, including the two strongly coupled magnetron modes. These tools will be invaluable in our current investigation of the important question of systematic errors. We are hopeful that by precisely controlling the motion of the ions and characterizing the electrostatic and magnetic fields they sample, we will be able to achieve mass comparison with a resolution approaching 1×10^{-11} in the near future.

Acknowledgments

This work is supported by the National Science Foundation.

References

1. R.S. Vandyck, D.L. Farnham, and P.B. Schwinberg: Phys. Rev. Lett. **70**, 2888 (1993)
2. R. Van Dyck, D. Farnham, S. Zafonte, and P. Schwinberg: 'High precision Penning trap mass spectroscopy and a new measurement of the proton's atomic mass'. In: *International Conference on Trapped Charged Particles and Fundamental Physics, Monterey, CA, August 31–September 4, 1998,* ed. by D.H.E. Dubin, D. Schneider (AIP, Woodbury, 1998) Vol. 457, pp. 101–110
3. R.S. Van Dyck, S.L. Zafonte, and P.B. Schwinberg: Hyperfine Interact. **132**, 163 (2001)
4. F. Difilippo, V. Natarajan, K.R. Boyce, and D.E. Pritchard: Phys. Rev. Lett. **73**, 1481 (1994)
5. M.P. Bradley et al.: Phys. Rev. Lett. **83**, 4510 (1999)
6. E.R. Cohen and B.N. Taylor: Rev. Mod. Phys. **59**, 1121 (1987)
7. T. Udem et al., Phys. Rev. Lett: **79**, 2646 (1997)
8. T. Udem, J. Reichert, R. Holzwarth, and T.W. Hänsch: Phys. Rev. Lett. **82**, 3568 (1999)
9. T. Beier et al.: Phys. Rev. Lett. **88**, 011603 (2002)

10. D.L. Farnham, R.S. Vandyck, and P.B. Schwinberg: Phys. Rev. Lett. **75**, 3598 (1995)
11. S. Battesti et al.: 'Measurement of h/M_{Rb} with Ultracold Atoms'. In: *Conference on Precision Electromagnetic Measurements*, ed. by U. Feller (IEEE, Ottawa, Canada, 2002) p. 308
12. S. Gupta, K. Dieckmann, Z. Hadzibabic, and D.E. Pritchard: Phys. Rev. Lett. **89**, 140401 (2002)
13. E.G. Kessler et al.: Nucl. Instrum. Methods **457**, 187 (2001)
14. G.L. Greene, M.S. Dewey, E.G. Kessler, and E. Fischbach: Phys. Rev. D **44**, R2216 (1991)
15. J. Bonn et al.: Phys. Atom. Nuclei **63**, 969 (2000)
16. V.M. Lobashev: Phys. Atom. Nuclei **63**, 962 (2000)
17. L.S. Brown and G. Gabrielse: Rev. Mod. Phys. **58**, 233 (1986)
18. R.M. Weisskoff et al.: J. Appl. Phys. **63**, 4599 (1988)
19. E.A. Cornell, R.M. Weisskoff, K.R. Boyce, and D.E. Pritchard: Phys. Rev. A **41**, 312 (1990)
20. E.A. Cornell et al.: Phys. Rev. Lett. **63**, 1674 (1989)
21. V. Natarajan, K.R. Boyce, F. Difilippo, and D.E. Pritchard: Phys. Rev. Lett. **71**, 1998 (1993)
22. P. Bevington and D. Robinson: *Data Reduction and Error Analysis for the Physical Sciences*, 2nd ed. (McGraw-Hill, Boston, 1992)
23. P. Huber: *Robust Statistics* (Wiley, New York 1981)
24. P.J. Lindstrom, W.G. Mallard, Eds., *NIST Chemistry WebBook, NIST Standard Reference Database Number 69* (National Institute of Standards and Technology, Gaithersburg, 2001) (http://webbook.nist.gov)
25. G. Audi and A.H. Wapstra: Nucl. Phys. A **595**, 409 (1995)
26. A. Wapstra and G. Audi: Nuc Phys A **432**, 1 (1985)
27. E.G. Kessler et al.: Phys. Lett. A **255**, 221 (1999)
28. E.A. Cornell, K.R. Boyce, D.L.K. Fygenson, and D.E. Pritchard: Phys. Rev. A **45**, 3049 (1992)
29. S. Rainville et al.: Hyp. Interact. **132**, 177 (2001)
30. V. Natarajan, F. Difilippo, and D.E. Pritchard: Phys. Rev. Lett. **74**, 2855 (1995)
31. F. Difilippo, V. Natarajan, K.R. Boyce, and D.E. Pritchard: Phys. Rev. Lett. **68**, 2859 (1992)
32. R. Forward: J. Appl. Phys. **50**, 1 (1979)

Current Status of the Problem of Cosmological Variability of Fundamental Physical Constants

D.A. Varshalovich[1], A.V. Ivanchik[1], A.V. Orlov[1], A.Y. Potekhin[1], and
P. Petitjean[2]

[1] Department of Theoretical Astrophysics, Ioffe Physical-Technical Institute,
   St.-Petersburg, 194021, Russia
[2] Institut d'Astrophysique de Paris – CNRS, Paris, France

Abstract. We review the current status of the problem of cosmological variability
of fundamental physical constants, provided by modern laboratory experiments, Oklo
phenomena analysis, and especially astronomical observations.

1   Introduction

The problem of the talk is one of the hot point of contemporary physics and
cosmology. Current theories of fundamental interactions (e.g., SUSY GUT, Su-
perstring theory) predict two kinds of variations of fundamental constants. First,
they state that the fundamental constants are "running constants" depend on
the energy transfer in particle interactions (Fig. 1). It is a result of radiation
corrections and vacuum polarization effects. It has been reliably confirmed in
high-energy accelerator experiments. For example, the fine-structure constant
$\alpha = e^2/\hbar c$ equals $1/137.036$ at low energies ($E \to 0$) and $1/128.896$ at energy
90 GeV [1]. Such "running" of the constants has to be taken into account for
consideration of very early Universe.

Second, the current theories predict that the *low-energy limits* of the fundamental constants can vary in the course of cosmological evolution and take
on different values at different points of space-time. Multidimensional theories
(Kaluza-Klein type, "p-brane" models, and others) predict variations of fundamental physical constants as a direct result of the cosmological evolution of the
extra-dimensional subspace. It means that the true constants of nature are defined in higher dimensions and their three-dimensional projections we observe
do not need to be constant. In several theories (e.g. Superstrings/M-theory), the
variations of the constants result from the cosmological evolution of the vacuum
state (a vacuum condensate of some scalar field or "Quintessence"). In addition,
a possible non-uniqueness of the vacuum state in different space-time regions
would allow constants to have different values in different places.

Clearly, experimental detection of a space-time variability of the fundamental constants would be a great step forward in understanding Nature. Note,
however, that a numerical value of any dimensional physical parameter depends
on arbitrary choice of physical units. In turn, there is no way to determine
the units in a remote space-time region other than through the fundamental
constants. Therefore it is meaningless to speak of a variation of a dimensional

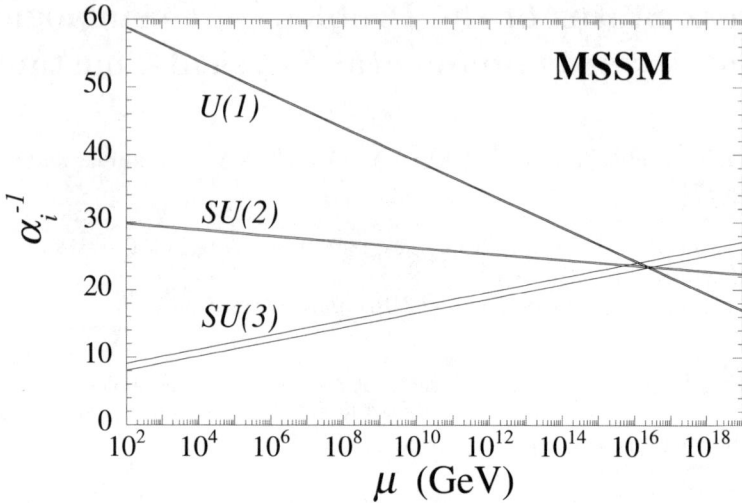

Fig. 1. Energy dependence of coupling constants in the Minimal Supersimmetric Standard Model of fundamental interactions

physical constant without specifying which of the other physical parameters are *defined* to be invariable. This point has been recently emphasized by Duff [2] (the counter-arguments by Moffat [3] are actually based on the unjustified implicit assumption that one can measure absolute time intervals in distant space-time regions without specifying the clock used). Usually, while speaking of variability of a dimensional physical parameter, one *implies* that *all* the other fundamental constants are fixed. So did Milne [4] and Dirac [5] in their pioneering papers devoted to a possible change of the gravitational constant G. More recently, a number of authors considered cosmological theories with a time varying speed of light c (e.g., Ref. [6] and references therein). However, if we adopt the standard definition of meter [7] as the length of path traveled by light in vacuum in $1/299\,792\,458$ s, then $c = 2.997\,924\,58 \times 10^{10}$ cm s^{-1} identically. Similarly, one cannot speak of variability of the electron mass m_e or charge e while using the Hartree units ($\hbar = e = m_e = 1$), most natural in atomic physics.

Thus, only *dimensionless* combinations of the physical parameters are truly fundamental, and only such combinations will be considered hereafter. We shall review the current status of the problem of space-time variability of the low-energy limits of the fundamental constants.

2   Tests for Possible Variations of Fundamental Constants

Various tests of the fundamental physical constants variability differ in space-time regions of the Universe which they cover (large review see e.g. [8]). In particular, *laboratory tests* infer the possible variation of certain combinations of constants "here and now" from comparison of different frequency standards. *Geophysical tests* impose constraints on combinations of fundamental constants over the past history of the Solar system, although most of these constraints are very indirect. In contrast, *astrophysical tests* (i.e. ones concerned with extragalactic observations) allows one to "measure" the values of fundamental constants in distant areas of the early Universe.

2.1 Local Tests

Laboratory Measurements. Laboratory tests are based on comparison of different frequency standards, depending on different combinations of the fundamental constants. Were these combinations changing differently, the frequency standards would eventually discord with each other. An interest in this possibility has been repeatedly excited since relative frequency drift was observed by several research groups using long term comparisons of different frequency standards. For instance, a comparison of frequencies of He-Ne/CH_4 lasers, NH_3 masers, H masers, and Hg^+ clocks with a Cs standard [9–13] has revealed relative drifts. Since the considered frequency standards have a different dependence on α via relativistic contributions of order α^2, the observed drift might be attributed to changing of the fine-structure constant. However, the more modern was the experiment, the smaller was the drift. Taking into account that the drift may be also related to some aging processes in experimental equipment, Prestage et al. [13] concluded that the current laboratory data provide only an upper limit $|\dot{\alpha}/\alpha| \leq 3.7 \times 10^{-14}$ yr^{-1}.

The most accurate experiment was performed recently by Sortais et al. [14]. They compared microwave clocks using laser cooled neutral atoms ^{87}Rb I and ^{133}Cs I. The frequencies of their HFS transitions of the ground states are

$$\nu_{\mathrm{HFS}}(^{87}\mathrm{Rb\,I}) = 6\ 834\ 682\ 610.904\ 343\ (17)\ \mathrm{Hz}$$

$$\nu_{\mathrm{HFS}}(^{133}\mathrm{Cs\,I}) = 9\ 192\ 631\ 770.000\ 000\ (0)\ \mathrm{Hz}$$

The error of $\nu_{\mathrm{HFS}}(^{133}\mathrm{Cs})$ equals zero by definition: it is the primary reference standard of frequency and the inverse time unit. Measurements of the ratio $\nu_{\mathrm{HFS}}(^{87}\mathrm{Rb})/\nu_{\mathrm{HFS}}(^{133}\mathrm{Cs})$ during 24 months indicate no change at the level of 3.1×10^{-15} yr^{-1}. This gives a new upper limit for α-variation $|\dot{\alpha}/\alpha| \leq 1.8 \times 10^{-14}$ yr^{-1}, providing the gyromagnetic ratios of ^{87}Rb and ^{133}Cs are invariable, either upper limit to the gyromagnetic ratio $|\dot{g}_p/g_p| \leq 4.2 \times 10^{-15}$ yr^{-1} providing the fine-structure constant is invariable.

Some other possibilities for laboratory searches of possible variations of the physical constants suggested by Karshenboim [15].

Analysis of the Oklo Phenomenon. The strongest limits to variation of the fine-structure constant α and the coupling constant of the strong interaction α_s have been originally inferred by Shlyakhter [16] from results of an analysis of the isotope ratio ^{149}Sm/^{147}Sm in the ore body of the Oklo site in Gabon, West Africa. This ratio turned out to be considerably lower than the standard one (instead of 0.92 it falls down to 0.006). It is believed to have occurred due to

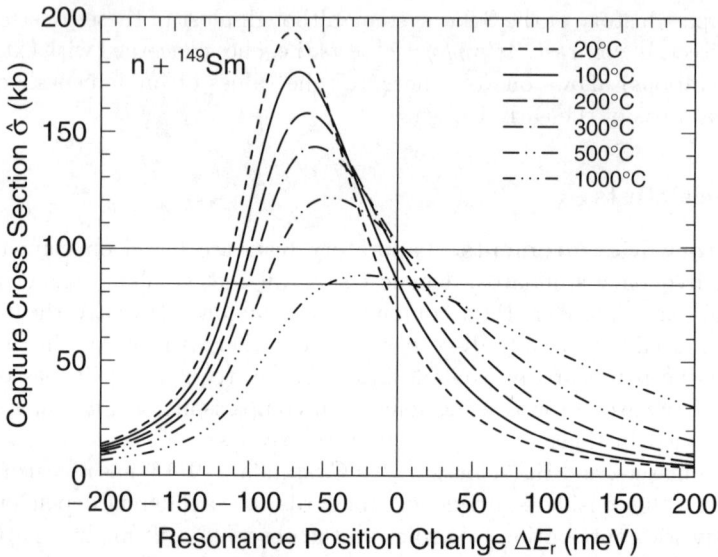

Fig. 2. Thermally averaged effective cross section for $n + {}^{149}Sm \rightarrow {}^{150}Sm + \gamma$ [18]. The horizontal two lines represent the range of the observed effective cross section (91 ± 6)kb

operation of the natural uranium fission reactor about 2×10^9 yr ago in those ores. One of the nuclear reactions accompanying this process was the resonance capture of neutrons by ${}^{149}Sm$ nuclei. Actually, the rate of the neutron capture reaction is sensitive to the energy of the relevant nuclear resonance level E_r (Fig. 2), which depends on the strong and electromagnetic interaction. Since the capture has been efficient 2×10^9 yr ago, it means that the position of the resonance has not shifted by more than its width ($\Gamma = 0.066$ eV) during the elapsed time. At variable α and invariable α_s (which is just a model assumption), the shift of the resonance level would be determined by changing the difference between the Coulomb energies of the ground-state nucleus ${}^{149}Sm$ and the nucleus ${}^{150}Sm^*$ excited to the level E_r. Unfortunately, there is no experimental data for the Coulomb energy of the excited ${}^{150}Sm^*$ in question. Using order-of-magnitude estimates, Shlyakhter [16] concluded that $|\dot{\alpha}/\alpha| \lesssim 10^{-17}$ yr^{-1}. From an opposite model assumption that α_s is changing whereas $\alpha =$ constant, he derived a bound $|\dot{\alpha}_s/\alpha_s| \lesssim 10^{-19}$ yr^{-1}.

Later Damour and Dyson [17] performed a more careful analysis, which resulted in the upper bound $|\dot{\alpha}/\alpha| \lesssim 7 \times 10^{-17}$ yr^{-1} (see, also Fujii et al., [19]). They have assumed that the Coulomb energy difference between the nuclear states of ${}^{149}Sm$ and ${}^{150}Sm^*$ in question is not less than that between the *ground* states of ${}^{149}Sm$ and ${}^{150}Sm$. The latter energy difference has been estimated from isotope shifts and equals ≈ 1 MeV. However, it looks unnatural that a weakly bound neutron (≈ 0.1 eV), captured by a ${}^{149}Sm$ nucleus to form the highly excited state ${}^{150}Sm^*$, can so strongly affect the Coulomb energy. Moreover, excited nuclei sometimes have Coulomb energies smaller than those for their ground states (e.g., Ref. [20]). This indicates the possibility of violation of the basic assumption involved in Ref. [17], and therefore this method may possess a lower actual sensitivity. Furthermore, a correlation between variations of α and α_s (which is likely in the frame of modern theory) might lead to considerable softening of the above-mentioned bound, as estimated by Sisterna and Vucetich [21].

Some Other Local Tests. Geophysical, geochemical, and paleontological data impose constraints on a possible changing of various combinations of fundamental constants over the past history of the Solar system, however most of these constraints are very indirect. A number of other methods are based on stellar and planetary models. The radii of the planets and stars and the reaction rates in them are influenced by values of the fundamental constants, which offers a possibility to check variability of the constants by studying, for example, lunar and Earth's secular accelerations. This was done using satellite data, tidal records, and ancient eclipses. Another possibility is offered by analyzing the data on binary pulsars and the luminosity of faint stars. Most of these have relatively low sensitivity. Their common weak point is the dependence on a model of a fairly complex phenomenon, involving many physical effects.

An analysis of natural long-lived α- and β-decayers in geological minerals and meteorites is much more sensitive. For instance, a strong bound, $|\dot{\alpha}/\alpha| < 5 \times 10^{-15}$ yr^{-1}, was obtained by Dyson [22] from an isotopic analysis of natural α- and β-decay products in Earth's ores and meteorites.

Having critically reviewed the wealth of the local tests, taking into account possible correlated synchronous changes of different physical constants, Sisterna and Vucetich [21] derived restrictions on possible variation rates of individual physical constants for ages t less than a few billion years ago, which correspond to cosmological redshifts $z \lesssim 0.2$. In particular, they have arrived at the estimate $\dot{\alpha}/\alpha = (-1.3 \pm 6.5) \times 10^{-16}$ yr^{-1}.

The most sensitive process is $^{187}Re \rightarrow ^{187}Os + e + \bar{\nu}$ due to a very small Q-value: $\Delta\tau/\tau \simeq 1.8 \times 10^4 \cdot \Delta\alpha/\alpha$. The laboratory measurement $\tau_{1/2}(lab) = (42.3 \pm 0.7)$ Gyr can be compared with the value inferred from Re/OS measurement in ancient meteorites $\tau_{1/2}(met) = (41.6 \pm 0.4)$ Gyr, dated by means of different radioactive methods (e.g. U/Th method, which is much less weakly affected by variation of α). The agreement within errors provides a significant constraint, $\Delta\alpha/\alpha = (1 \pm 1) \times 10^{-6}$ [23].

All the local methods listed above give estimates for only a narrow space-time region around the Solar system. For example, the epoch of the Oklo reactor $(1.8 \times 10^9$ years ago) corresponds to the cosmological redshift $z \approx 0.1$.

2.2 Quasar Spectra

Values of the physical constants in the early epochs are estimated directly from observations of quasars (the most powerful sources of radiation) whose spectra were formed when the Universe was several times younger than now.

The wavelengths of the spectral lines observed in radiation from these objects (λ_{obs}) increase compared with the laboratory values (λ_{lab}) in proportion $\lambda_{\text{obs}} = \lambda_{\text{lab}}(1 + z)$, where z is the *cosmological redshift* which can be used to determine the age of the Universe at the line-formation epoch. Analyzing these spectra we can study the epochs when the Universe was several times younger than now.

At present, the extragalactic spectroscopy enables one to probe the physical conditions in the Universe up to cosmological redshifts $z \lesssim 6$, which correspond, by order of magnitude, to the scales $\lesssim 15$ Gyr in time and $\lesssim 5$ Gpc in space. The large time span enables us to obtain quite stringent estimates of the rate of possible time variations, even though the astronomical wavelength measurements are not so accurate as the precision metrological experiments. Moreover, such analysis allows us to study the physical conditions in distant regions of the Universe, which were causally disconnected at the line-formation epoch.

In general, the dependence of wavelengths of resonant lines in quasar spectra on fundamental constants is not the same for different transitions. This makes it possible to distinguish the cosmological redshift (common for all lines in a given absorption system) from the shift due to the possible variation of fundamental constants.

Fine-Structure Constant. Quasar spectra were used for setting bounds on possible variation rates of fundamental physical constants by many authors. The first ones were Bahcall *et al.* [24,25], who compared the observed redshifts z of the components of fine-structure doublets in spectra of distant quasars, and derived the estimates $\Delta\alpha/\alpha = (-2\pm5) \times 10^{-2}$ at $z = 1.95$ and $\Delta\alpha/\alpha = (-1\pm2) \times 10^{-3}$ at $z = 0.2$. Afterwards this and similar methods were used for setting stronger bounds on $\Delta\alpha/\alpha$ at different z. In particular, Potekhin and Varshalovich [26] applied modern statistical methods to analysis of ≈ 1400 pairs of wavelengths of the fine-splitted doublet absorption lines in quasar spectra and obtained an upper bound on the rate of a relative variation of the fine-structure constant $|\alpha^{-1}d\alpha/dz| < 5.6 \times 10^{-4}$ for the epoch $0.2 \le z \lesssim 4$. Later we (Ivanchik *et al.* [27]) optimized the strategy of studying the time-dependence of α. As a result, a new constraint on the possible deviation of the fine-structure constant at $z = 2.8$–3.1 from its present ($z = 0$) value was obtained: $|\Delta\alpha/\alpha| < 1.6 \times 10^{-4}$. The corresponding upper limit of the α variation rate averaged over $\sim 10^{10}$ yr is $|\dot\alpha/\alpha| < 2 \times 10^{-14}$ yr^{-1}.

In a recent series of papers, Webb *et al.* (e.g., Ref. [28] and references therein) reported a possible detection of variation of the fine-structure constant, $\Delta\alpha/\alpha = -0.72 \pm 0.18 \times 10^{-5}$, averaged over the cosmological redshifts $z = 0.5$–3.5. However, it is difficult to evaluate systematic errors which might simulate this result. In particular, the method used by the authors, which is based on simultaneous measurements of wavelengths of a large number of transitions for various ions, depends more sensitively on poorly known factors (e.g., isotope variations, instrumental calibration errors, etc.) than in the method based on separate measurements of fine structure of spectral lines of each species [26,27]. On the other hand, the latter method has a larger statistical error than the method of Webb *et al.* Therefore, it is especially important to check possible variations of different fundamental constants, using different techniques, applied to different stages of the cosmological evolution.

Proton-to-Electron Mass Ratio. Since any interaction inherent in a given particle contributes to its observed mass, a variation in α suggests a variation in the proton-to-electron mass ratio $\mu = m_p/m_e$. The functional dependence $\mu(\alpha)$ is currently unknown, but there are several theoretical models which allow one to estimate the electromagnetic contribution to μ (e.g., [29,30]), as well as model relations between cosmological variations of α and μ [31].

Evaluation of μ in distant space-time regions of the Universe is possible in quasar spectra. The wavelengths of these lines depend on μ through the reduced mass of the molecule. The method is based on the relation [32]

$$\frac{1+z_i}{1+z_k} = \frac{(\lambda_i/\lambda_k)_z}{(\lambda_i/\lambda_k)_0} \simeq 1 + (K_i - K_k)\left(\frac{\Delta\mu}{\mu}\right), \tag{1}$$

where z_i is the observed redshift of an individual line, the subscripts 'z' and '0' mark the wavelength ratios in the quasar spectrum and the terrestrial laboratory, respectively, and $K_i \equiv \partial\ln\lambda_i/\partial\ln\mu$ are the *sensitivity coefficients*. A method for calculation these coefficients has been presented in Ref. [33]. The authors have applied a linear regression (z as a linear function of K) analysis to the H_2 absorption lines in the spectrum of quasar PKS 0528−250 at $z = 2.8108$ and obtained an estimate of the fractional variation of $\Delta\mu/\mu = (-11.5 \pm 7.6) \times 10^{-5}$. Thus, no statistically significant variation was found. The above estimate approximately corresponds to the upper bound $|\dot\mu/\mu| < 1.5 \times 10^{-14}$ yr^{-1}.

Recently, similar analyses of the H_2 absorption system in the spectrum of quasar Q 0347−382 at $z = 3.0249$ have been performed by Levshakov *et al.* [34] and Ivanchik *et al.* [35]; the latter authors analyzed also the H_2 absorption system in the spectrum of quasar Q 1232+082 at $z = 2.3377$. The most conservative estimate for the possible variation of μ in the past ~ 10 Gyr, obtained in Ref. [35], reads

$$\Delta\mu/\mu = (5.7 \pm 3.8) \times 10^{-5}. \tag{2}$$

The corresponding linear regression is illustrated in Fig. 3. Thus, we have obtained the most stringent estimate on a possible cosmological variation of mu $|\dot\mu/\mu| < 6 \times 10^{-15}$ yr^{-1}.

2.3 Cosmic Microwave Background Radiation

Any time variation in the fine-structure constant (as well as m_e) alters the ionization history of the Universe and therefore changes the pattern of cosmic microwave background fluctuations (Fig. 4). Changing α changes the energy levels of hydrogen, the Thomson cross section, and recombination rates. These changes are dominated by the change in the redshift of recombination due to the shift in the binding energy of hydrogen.

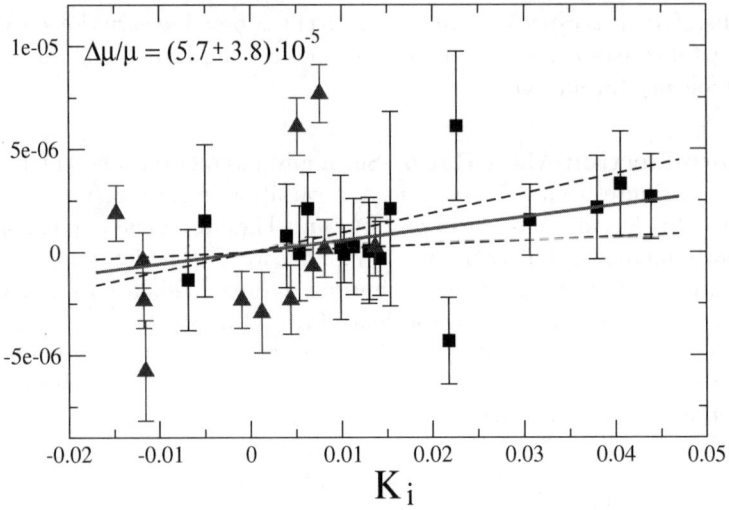

Fig. 3. Regression analysis of ξ_i-to-K_i for the H_2 lines. $\xi_i = (z_i - \overline{z})/(1 + \overline{z})$

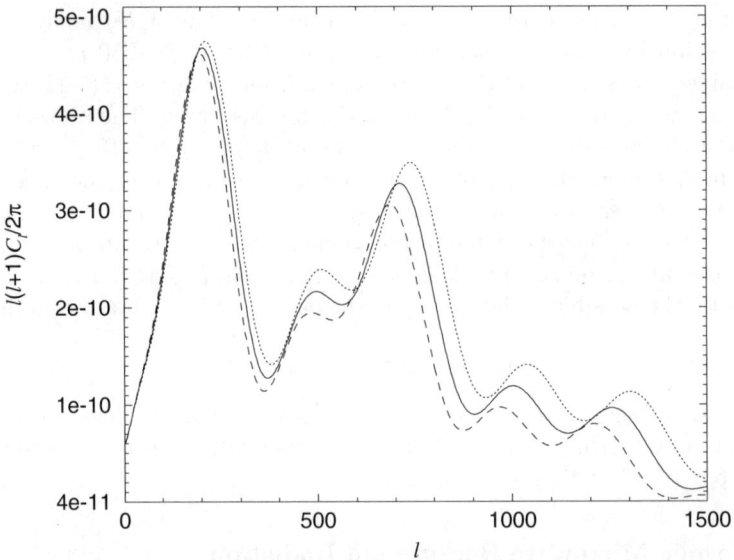

Fig. 4. The spectrum of CMB fluctuations for the standard scenario (SCDM, $\Omega_b = 0.05$, $h = 0.65$) (solid curve), an increase of α by 3% (dotted curve), and a decrease of α by 3% (dashed curve)

An analysis of the recently obtained data from BOOMERanG [36] and MAX-IMA [37] experiments allowing for the possibility of a time-varying the fine-structure constant. This data prefers a value of α that was smaller in the past (which is in agreement with measurements of α from quasar observations). However, the strong statements about α can not be made because such a theoretical analysis involves several additional parameters (cosmological ones H_0, Ω_0, and Ω_b as well as some physical constants, e.g. the electron mass). Bounds imposed on the variation of α can be significantly relaxed if one also allows for a change in the equation of state of quintessence which mimic the cosmological Λ-term. In any case, such a analysis allows to obtained upper limit on α-variation at the recombination epoch [38–41]:

$$|\Delta\alpha/\alpha| \leq 10^{-2} . \tag{3}$$

2.4   Primordial Nucleosynthesis

Temporal variations of the coupling constants can be revealed by analyzing the dependence of the primordial ^{4}He mass fraction on gravitational constant G, on fine-structure constant α, and on other constants [42]. Based on astronomical observational data for the primordial helium abundance Y_p, Kolb et al. [42] imposed constraints (2-3%) on the possible deviations of fundamental physical constants at the epoch of primordial nucleosynthesis from their current values. However, they varied different constants separately (with the remaining constants being fixed) and assumed the parameter $\eta = n_B/n_\gamma$, the baryonic-to-photon density ratio, to be also fixed.

Subsequently, modifying the standard nucleosynthesis theory, several authors imposed constraints on the relative change in fundamental physical constants by taking into account the possible simultaneous change in various constants (see, e.g., [43–45]). However, the calculations were performed for a specific fixed η (as in the pioneering study by Kolb et al. [42]).

In paper [46] a two-parameter (η, δ) model for primordial nucleosynthesis was considered, in which η is a free parameter and can depend (for a constant yield of light elements) on the deviation of constant δ. The parameter δ characterizes the relative deviation of fundamental physical constants at the epoch of primordial nucleosynthesis from their current values (in particular $\delta = \Delta\alpha/\alpha$).

Unfortunately, the Primordial Nucleosynthesis as well as Cosmic Microwave Background Radiation do not give very stringent limitation on the variation of fundamental constant because of many different parameters involved in the analysis.

3   Conclusions

We have discussed the current status of the problem of cosmological variability of fundamental physical constants, making emphasis on the studies of the space-time variability of two basic parameters of atomic and molecular physics: the fine-structure constant α and the proton-to-electron mass ratio μ. A variation of these parameters is not firmly established. More precise measurements and observations and their accurate statistical analyses are required in order to detect the expected variations of the fundamental constants.

Acknowledgments

This work has been supported in part by RFBR (02-02-16278a, 03-02-06279-mac) and the RAS Program "Non-stationary phenomena in astronomy".

References

1. F. Jegerlehner: Nuclear Physics B (Proc. Suppl.) **51C**, 131 (1996)
2. M.J. Duff: hep-th/0208093
3. J.W. Moffat: hep-th/0208109
4. E. Milne: Proc. R. Soc. London A**158**, 324 (1937)
5. P.A.M. Dirac: Nature **139**, 323 (1937); Proc. R. Soc. London A**165**, 199 (1938)
6. J.W. Moffat: astro-ph/0210042
7. B.W. Petley: Nature **303**, 373 (1983)
8. J.-P. Uzan: hep-ph/0205340
9. N.I. Kolosnitsin, S.B. Pushkin, and V.M. Purto: Problems of gravitational theory and elementary particles **7**, 208 (1976)
10. Yu.S. Domnin, A.N. Malimon, V.M. Tatarenkov, P.S. Schumyanskii: Pisma v Zhurn. Eksp. Teor. Fiz. **43**, 167 (1986); [JETP Lett. **43**, 212 (1986)]
11. N.A. Demidov, E.M. Ezhov, B.A. Sakharov, B.A. Uljanov, A. Bauch, and B. Fisher: in *Proc. of 6th European Frequency and Time Forum*, European Space Agency, p. 409 (1992)
12. L.A. Breakiron: in *Proc. 25th Annual Precise Time Interval Applications and Planning Meeting*, NASA Conference Publication No. 3267, p. 401 (1993)
13. J.D. Prestage, R.L. Tjoelker, and L. Maleki: Phys. Rev. Lett. **74**, 3511 (1995)
14. Y. Sortais, S. Bize, M. Abgrall, S. Zhang, C. Nicolas, C. Mandache, P. Lemonde, P. Laurent, G. Santarelli, N. Dimarcq, P. Petit, A. Clairon, A. Mann, A. Luiten, S. Chang, and C. Salomon: Physica Scripta **T95**, 50 (2001)
15. S.G. Karshenboim: Can. J. Phys. **78**, 639 (2002)
16. A.I. Shlyakhter: Nature **25**, 340 (1976)
17. T. Damour and F.J. Dyson: Nucl. Phys. B **480**, 37 (1996)
18. Y. Fujii, A. Iwamoto, T. Fukahori, T. Ohnuki, M. Nakagawa, H. Hidaka, Y. Oura and P. Möller: Nucl. Phys. B **573**, 377 (2000)
19. Y. Fujii, A. Iwamoto, T. Fukahori, T. Ohnuki, M. Nakagawa, H. Hidaka, Y. Oura, and P. Möller: hep-ph/0205206
20. G.M. Kalvius and G.K. Shenoy: Atomic and Nuclear Data Tables **14**, 639 (1974)
21. P.D. Sisterna and H. Vucetich: Phys. Rev. D **41**, 1034 (1990)
22. F.J. Dyson: in *Aspects of Quantum Theory*, eds. A. Salam, E.P. Wigner, Cambridge Univ. Press, Cambridge, p. 213 (1972)
23. G. Fiorentini and B. Ricci: astro-ph/0207390
24. J. Bahcall and M. Schmidt: Phys. Rev. Lett. **19**, 1294 (1967)
25. J. Bahcall, W.L.W. Sargent, and M. Schmidt: Astrophys. J. **149**, L11 (1967)
26. A.Y. Potekhin and D.A. Varshalovich: Astron. Astrophys. Suppl. Ser. **104**, 89 (1994)

27. A.V. Ivanchik, A.Y. Potekhin, and D.A. Varshalovich: Astron. Astrophys. **343**, 439 (1999)
28. J.K. Webb, M.T. Murphy, V.V. Flambaum, V.A. Dzuba, J.D. Barrow, C.W. Churchill, J.X. Prochaska, and A.M. Wolfe: Phys. Rev. Lett. **87**, 091301 (2001)
29. J. Gasser and H. Leutwyler: Phys, Rep. **87**, 77 (1982)
30. T. Damour and A.M. Polyakov: Nucl. Phys. B **423**, 532 (1994)
31. X. Calmet and H. Fritzsch: Eur. Phys. J. **C24**, 639 (2002); /hep-ph/0112110
32. D.A. Varshalovich and S.A. Levshakov: Pis'ma v Zhurn. Eksper. Teor. Fiz. **58**, 231 (1993); [JETP Lett. **43**, 212 (1986)]
33. A.Y. Potekhin, A.V. Ivanchik, D.A. Varshalovich, K.M. Lanzetta, J.A. Baldwin, G.M. Williger, and R.F. Carswell: Astrophys. J. **505**, 523 (1998)
34. S.A. Levshakov, M. Dessauges-Zavadsky, S. D'Odorico, and P. Molaro, Mon. Not. Roy. Astron. Soc. **333**, 373 (2002).
35. A.V. Ivanchik, E. Rodriguez, P. Petitjean, and D.A. Varshalovich: Astronomy Lett. **28**, 423 (2002)
36. P. de Bernardis et al.: Nature **404**, 939 (2000)
37. S. Hanany et al.: Astrophys. J. **545**, L5 (2000)
38. S. Hannestad: Phys. Rev. D **60**, 023515 (1999)
39. M. Kaplinghat, R.J. Scherrer, and M.S. Turner: Phys. Rev. D **60**, 023516 (1999)
40. P.P. Avelino, S. Esposito, G. Mangano, C.J.A.P. Martins, A. Melchiorri, G. Miele, O. Pisanti, G. Rocha, P.T.P. Viana: Phys. Rev. D **64**, 103505 (2001)
41. G. Huey, S. Alexander, L. Pogosian: Phys. Rev. D **65**, 083001 (2002)
42. E.W. Kolb, M.J. Perry, and T.P. Walker: Phys. Rev. D **33**, 869 (1986)
43. B.A. Campbell and K.A. Olive: Phys. Lett. B **345**, 429 (1995)
44. D.I. Santiago, D. Kalligas, and R.V. Wagoner: Phys. Rev. D **56**, 7627 (1997)
45. L. Bergstrom, S. Iguri, and H. Rubinstein: Phys. Rev. D **60**, 045005 (1999)
46. A.V. Ivanchik, A.V. Orlov, and D.A. Varshalovich: Astronomy Lett. **27**, 615 (2001)

Appendix:
Proceedings of International Conference on
Precision Physics of Simple Atomic Systems
(St. Petersburg, 2002)

Canadian Journal of Physics **80**(11) (2002)

Table of Contents

Nonresonant corrections for the hydrogen atom
L.N. Labzowsky, D.A. Solovyev, G. Plunien, and G. Soff 1187

Progress in helium fine-structure calculations and the
fine-structure constant
G.W.F. Drake 1195

Two-loop QED bound-state calculations and squared decay rates
U.D. Jentschura, Ch.H. Keitel, and K. Pachucki 1213

Optical measurement of the 2S hyperfine interval in atomic hydrogen
M. Fischer, N. Kolachevsky, S.G. Karshenboim, and T.W. Hänsch 1225

Measurement of the g_J factor of a bound electron in hydrogen-like
oxygen $^{16}O^{7+}$
*J.L. Verdú, S. Djekic, T. Valenzuela, H. Häffner, W. Quint, H.J. Kluge,
and G. Werth* 1233

New value for the mass of the electron from an experiment on the g factor
in $^{12}C^{5+}$ and $^{16}O^{7+}$
*T. Beier, S. Djekic, H. Häffner, N. Hermanspahn, H.-J. Kluge, W. Quint,
S. Stahl, T. Valenzuela, J. Verdú, and G. Werth* 1241

One-loop self-energy correction to the bound-electron g factor
V.A. Yerokhin, P. Indelicato, and V.M. Shabaev 1249

Testing of QED-theory on the Rydberg series for the He-like
multicharged ions
V.G. Pal'chikov, I.Yu. Skobelev and A.Ya. Faenov 1255

Hyperfine quenching of the $2^3P_{0,2}$ states in He-like ions
A.V. Volotka, V.M. Shabaev, G. Plunien, G. Soff, and V.A. Yerokhin .. 1263

Precision physics with light muonic and hadronic atoms
V.E. Markushin 1271

Search for an exotic three-body decay of orthopositronium $o - Ps \rightarrow \gamma + X_1 + X_2$
P. Crivelli 1281

A Detector with high-detection efficiency in 4- and 5-photon-positronium annihilations
M. Chiba, J. Nakagawa, H. Tsugawa, R. Ogata, and T. Nishimura 1287

Hadronic effects in leptonic systems: muonium hyperfine structure and anomalous magnetic moment of muon
S.I. Eidelman, S.G. Karshenboim and V.A. Shelyuto 1297

g factor in a light two body atomic system: a determination of fundamental constants to test QED
S.G. Karshenboim and V.G. Ivanov 1305

Development of the new method of the positronium generation. Abilities and future trends
V. Antropov, A. Ivanov, Yu. Korotaev, T. Mamedov, I. Meshkov,
I. Seleznev, A. Sidorin, A. Smirnov, E. Syresin, G. Trubnikov,
and S. Yakovenko 1313

VNIIFTRI Cesium Fountain
Yu. S. Domnin, G.A. Elkin, A.V. Novoselov, L.N. Kopylov,
V.N. Baryshev, and V.G. Pal'chikov 1321

Single ion mass spectrometry at 100 ppt and beyond
S. Rainville, J.K. Thompson, D.E. Pritchard 1329

The structure of light nuclei and its effect on precise atomic measurements
J.L. Friar 1337

Systematic model calculations of HFS in light and heavy ions
M. Tomaselli, Th. Kühl, W. Nörtershäuser, G. Ewald, R. Sanchez,
S. Fritzsche, and S.G. Karshenboim 1347

Resent results and current status of the muon $g - 2$ experiment at BNL
S. Redin, R.M. Carey, E. Efstathiadis, M.F. Hare, X. Huang, F. Krinen,
A. Lam, J.P. Miller, J. Paley, Q. Peng, O. Rind, B.L. Roberts,
L.R. Sulak, A. Trofimov, G.W. Bennett, H.N. Brown, G. Bunce,
G.T. Danby, R. Larsen, Y.Y. Lee, W. Meng, J. Mi, W.M. Morse,
D. Nikas, C. Ozben, R. Prigl, Y.K. Semertzidis, D. Warburton,
V.P. Druzhinin, G.V. Fedotovich, D. Grigoriev, B.I. Khazin,
I.B. Logashenko, N.M. Ryskulov, Yu.M. Shatunov, E.P. Solodov,
Yu.F. Orlov, D. Winn, A. Grossmann, K. Jungmann, G. zu Putlitz,
P. von Walter, P.T. Debevec, W. Deninger, F. Gray, D.W. Hertzog,
C.J.G. Onderwater, C. Polly, S. Sedykh, M. Sossong, D. Urner,
A. Yamamoto, B. Bousquet, P. Cushman, L. Duong, S. Giron,
J. Kindem, I. Kronkvist, R. McNabb, T. Qian, P. Shagin,
C. Timmermans, D. Zimmerman, M. Iwasaki, M. Kawamura, M. Deile,

H. Deng, S.K. Dhawan, F.J.M. Farley, M. Grosse-Perdekamp,
V.W. Hughes, D. Kawall, J. Pretz, E.P. Sichtermann,
and A. Steinmetz 1355

Gyromagnetic factors of bound particles with arbitrary spin in quantum
electrodynamics
R.N. Faustov and A.P. Martynenko 1365

Precise energies of highly excited hydrogen and deuterium
S. Kotochigova, P.J. Mohr, and B.N. Taylor 1373

Paschen-Back effect in helium spectra revisited
V.D. Ovsiannikov and E.V. Tchaplyguine 1381

Diamagnetic effect on the intensity of helium radiation lines
V.D. Ovsiannikov and V.V. Chernushkin 1391

Precise theory of the Stark effect on hydrogen- and helium-like atoms
V.D. Ovsiannikov and V.G. Pal'chikov 1401

Relativistic linear response wave function of the lowest $ns_{1/2}$-states in
hydrogen-like atoms. New analytic results
V. Yakhontov 1413

Non-variational and non-adiabatic calculations on the hydrogen molecular
ion and its μ^--isotopes
V. Yakhontov and Jungen 1423

Index

Accurate nuclear calculations
- $A = 2\text{-}10$, 65
- ^{3}H and ^{3}He, 65
- ^{4}He, 65
Alkali atoms, 178
Anomalous magnetic moment, *see* g factor
Antiprotonic hydrogen, 52
Atomic cascade, 38
Atomic mass
- hydrogen, 186
- neutron, 186
- deuterium, 186
- carbon, 186
- nitrogen, 186
- oxygen, 186
- neon, 186
- sodium, 186
- silicon, 186
- argon, 186
- rubidium, 186
- cesium, 186
Avogadro constant, 178

Chemical binding energies, 180
Chiral perturbation theory, 61, 64
Chiral symmetry
- in QCD, 60
- in QED, 60
Compton amplitude
- meson-exchange currents, 70
- nuclear, 70
Coulomb Green function, 15, 19–23, 26–28
Coulomb deexcitation, 40

Determination of
- the fine structure constant, 6, 155–156, 178
- the mass of electron, 5
- the muon-to-electron mass ratio, 152
- the proton-to-electron mass ratio, 154
- the Rydberg constant, *see* Rydberg constant

Electric polarizability, 68
- Coulomb corrections, 68
- deuterium, 67, 75
- light nuclei, 71
- tensor, 69
Electron-nucleus scattering, 69, 73
- Coulomb corrections, 73
Electronic refrigeration, 194
Exotic atoms, 7–8
- pionic hydrogen, *see* Pionic hydrogen
- kaonic hydrogen, *see* Kaonic hydrogen
Extended standard cascade model, 41
External Auger effect, 40

Gerasimov-Drell-Hearn sum rule, 73
Green's function Monte Carlo, 65
g factor
- of electron
- - in muonium, 150–152
- - in hydrogen-like ion, 152–155
- - in hydrogen-like ion, contribution, 105–111
- - in hydrogen-like carbon, 154–155
- - in hydrogen-like oxygen, 154–155
- of muon
- - in muonium, 150–152
- - free, 163–174

Hyperfine splitting
- deuterium, 71, 73
- light atoms, 72
- quadrupole, 69
- spin-dependent polarizability, 71, 72
- Zemach moments, 72, 73

Hyperfine structure
– in positronium, 149–150
– in muonium, 148–150
– in hydrogen, 145–147
– in deuterium, 145–147
– in tritium, 145–146
– in helium-3 ion, 145–147
– in hydrogen-like ion
– – contribution, 105–111
– in lithium-like ion
– – contribution, 111–112
– D_{21}, 2–3, 147

Ion trap, 180
Isotope shift
– hydrogen isotopes, 74
– deuterium-hydrogen, 75
– tritium-hydrogen, 76
– helium isotopes, 74
– lithium isotopes, 74
– light atoms, 74

Johnson noise, 193

Kaonic hydrogen, 49

Lamb shift
– in hydrogen, 23, 25, 142–144
– in muonic hydrogen, 42, 73
– in muonic ^{4}He, 71
– nuclear finite size, 67, 68, 73
Lyman-α transition, 26–28

Magnetic moment
– nuclear, 69
Magnetic susceptibility
– deuterium, 75
Microwave background radiation, 205
Mode coupling in trap, 183
Muonic atoms, 7
Muonic hydrogen
– Lamb shift, see Lamb shift in muonic
  hydrogen
– metastable $2S$ state, 43

Neutrino mass, 180
Neutron capture, 178
Nonresonant corrections, 25–28
Nuclear charge density, 69, 71
– exotic contributions, 74

Nuclear current density, 69
– magnetization density, 71
– meson-exchange currents, 72
Nuclear force
– first-generation, 64
– long-range, 61
– OPEP, 62, 64
– second-generation, 64
– short-range, 61, 62, 64
– tensor, 64
– third-generation, 64
– three-nucleon, 62, 64, 65
– TPEP, 62, 64
Nuclear phenomenology, 61, 63
Nuclear quadrupole moment
– Coulomb interaction, 69
– deuterium, 69
Nuclear radii, 69
– proton, 73, 74
– neutron, 74, 75
– deuterium, 75
– tritium, 70, 75
– ^{3}He, 79
– magnetic, 69
– wave function, 74, 75
Nuclear scales, 67, 71
Nuclear shifts of atomic levels, 143–146

π-pulse, 183
Penning trap, 180
Pionic hydrogen, 46
Polarizability
– electric, 67, 71
– generalized, 71
– logarithmic modification, 71
– magnetic, 71
– spin-dependent, 71
Positronium
– rare decay, 3
Power counting, 61, 67, 72
Proton charge radius, 42, 143–144

QCD, 60
QED, 141–158
– tests, 6
Quantum electrodynamics, see QED

Radiative deexcitation, 38
Reproduction of kilogram, 178
Rydberg constant, 142–144

Simple atoms, 1–9, 141
Single ion, 180
Squeezed states, 193
SQUID, 182
Standard cascade model, 41
Stark mixing, 39
Sturmian expansion, 15, 18, 20
Sturmian function, 18, 19
Subthermal detection, 193

Test of special relativity, 179

Vacuum polarization
– hadronic, 69

Variations of the fundamental constants,
    6, 199–209
– α, 201, 203–205, 207
– electron-to-proton mass ratio, 205
– geochemical search, 201–203
– laboratory experiments, 201
– primordial nucleasynthesis, 207
– proton g factor, 201
– quasar spectra, 203–205
Virial relations, 97–113

Whittaker function, 17

$Z\alpha$ expansion, 25–28, 33

Druck: Strauss Offsetdruck, Mörlenbach
Verarbeitung: Schäffer, Grünstadt